FUNDAMENTALS OF ENGINEERING
INDUSTRIAL DISCIPLINE
SAMPLE QUESTIONS & SOLUTIONS

FUNDAMENTALS OF ENGINEERING

INDUSTRIAL DISCIPLINE
SAMPLE QUESTIONS & SOLUTIONS

Published by the
National Council of Examiners for Engineering and Surveying®
280 Seneca Creek Road, Clemson, SC 29631 800-250-3196 www.ncees.org

ISBN-13: 978-1-932613-28-5
ISBN-10: 1-932613-28-5

Printed in the United States of America

TABLE OF CONTENTS

UPDATES TO EXAMINATION INFORMATION

For current exam specifications, errata for this book, a list of calculators that may be used at the exam, guidelines for requesting special accommodations, and other information about exams, visit the NCEES Web site at www.ncees.org.

INTRODUCTION

One of the functions of the National Council of Examiners for Engineering and Surveying (NCEES) is to develop examinations that are taken by candidates for licensure as professional engineers. NCEES has prepared this handbook to assist candidates who are preparing for the Fundamentals of Engineering (FE) examination in general engineering. NCEES is an organization established to assist and support the licensing boards that exist in all states and U.S. territories. NCEES provides these boards with uniform examinations that are valid measures of minimum competency related to the practice of engineering.

To develop reliable and valid examinations, NCEES employs procedures using the guidelines established in the *Standards for Educational and Psychological Testing* published by the American Psychological Association. These procedures are intended to maximize the fairness and quality of the examinations. To ensure that the procedures are followed, NCEES uses experienced testing specialists possessing the necessary expertise to guide the development of examinations using current testing techniques.

The examinations are prepared by committees composed of professional engineers from throughout the nation. These engineers supply the content expertise that is essential in developing examinations that are valid measures of minimum competency.

LICENSURE: AN IMPORTANT DECISION

One of the most important decisions you can make early in your engineering career is to place yourself on a professional course and plan to become licensed as a professional engineer (P.E.). The licensure of professional engineers is important to the public because of the significant role engineering plays in society. The profession regulates itself, through the licensing boards, by setting high standards for professional engineers. These high standards help protect the public by requiring that professional engineers demonstrate their competence to practice in a manner that will safeguard the public's safety and welfare.

The first step on the path to licensure as a P.E. is to take and pass the FE examination. If you are a student or a recent college graduate, you are well advised to take this step while coursework is still fresh in your mind. After passing the examination, your state board will designate you as an engineer intern (EI). In the past, the term "engineer-in-training" (EIT) has been used to recognize this step in your career.

To continue the licensure process, typically you must complete 4 years of progressive and verifiable experience that is acceptable to your licensing board. Some boards require that this experience be gained under the supervision of a professional engineer. Again, because of variations from state to state, you should contact your licensing board for information to ensure that you are on track to meet their requirements. Having met these requirements, you will be granted permission to take the Principles and Practice of Engineering (PE) examination. After you pass the PE examination, you may become licensed as a professional engineer and use the distinguished P.E. designation.

Licensing Requirements

Eligibility
The primary purpose of licensure is to protect the public by evaluating the qualifications of candidates seeking licensure. While examinations offer one means of measuring the competency levels of candidates, most licensing boards also screen candidates based on education and experience requirements. Because these requirements vary between boards, it would be wise to contact the appropriate board. Board addresses and telephone numbers may be obtained by visiting our Web site at www.ncees.org or by calling 800-250-3196.

Application Procedures and Deadlines
Application procedures for the examination and instructional information are available from individual boards. Requirements and fees vary among the boards, and applicants are responsible for contacting their board office. Sufficient time must be allotted to complete the application process and assemble required data.

Description of Examinations

Examination Schedule
The NCEES FE examination is offered to the boards in the spring and fall of each year. Dates of future administrations are published on the NCEES Web site at www.ncees.org. You should contact your board for specific locations of exam sites.

Examination Content
The purpose of the FE examination is to determine if the examinee has an adequate understanding of basic science, mathematics, engineering science, engineering economics, and discipline-specific subjects normally covered in coursework taken in the last 2 years of an engineering bachelor degree program. The examination identifies those applicants who have demonstrated an acceptable level of competence in these subjects.

The 8-hour FE examination is a no-choice examination in a multiple-choice format. The examination is administered in two 4-hour sessions. The morning session contains 120 questions, and the afternoon session contains 60 questions. Each question has four answer options. The examination specifications presented in this book give details of the subjects covered on the examinations.

Numerical questions are posed mostly in metric units, normally International System of Units (SI). However, because some subjects are typically taught in U.S. customary units (in.-lb) only, questions on the examination dealing with these subject areas are posed in U.S. customary units.

The FE examination is a closed-book examination. However, since engineers rely heavily on reference materials, you will be given a copy of the NCEES *FE Supplied-Reference Handbook* at the examination site. The *Handbook* contains formulas and data that examinees cannot reasonably be expected to commit to memory. The *Handbook* does not contain all the information required to answer every question on the examination. For example, basic theories, formulas, and definitions that examinees are expected to know have not been included. To familiarize yourself with the content of the *Handbook* before the examination, visit the NCEES Web site at www.ncees.org to view and download a free copy of the *Handbook*. You may also order, at minimal cost, a hard copy of the *Handbook* on the NCEES Web site or by calling NCEES Customer Service at 800-250-3196. You will not be allowed to take your copy of the *Handbook*

into the examination; you must use the copy provided to you by the proctor in the examination room.
A sample examination is presented in this book. By illustrating the general content of the subject areas and formats, the questions should be helpful in preparing for the examination. Solutions are presented for all the questions. The solution presented may not be the only way to solve the question. The intent is to demonstrate the typical effort required to solve each question.

No representation is made or intended as to future examination questions, content, or subject matter.

Examination Preparation and Review
Examination development and review workshops are conducted at least twice annually by standing committees of the NCEES. Additionally, workshops are held as required to supplement the bank of questions available. The content and format of the questions are reviewed by the committee members for compliance with the specifications and to ensure the quality and fairness of the examination. These engineers are selected with the objective that they be representative of the profession in terms of geography, ethnic background, gender, and area of practice.

SCORING OF THE EXAMINATION

Both sessions of the FE examination are worth the same total number of points, and questions are weighted at one point for each morning question and two points for each afternoon question. Within each session, every question has equal weight. Your final score on the examination is arrived at by the summation of the numbers of points obtained in each session.

The score required to pass the examination varies between administrations of the examination. This acknowledges the fact that variations in difficulty may exist between different versions of the examination. The individual passing scores are set to reflect a minimum level of competency consistent with the purpose of licensing examinations. This procedure assures candidates that they have the same chance of passing the examination even though the difficulty of an examination may vary from one administration to another.

To accomplish the setting of fair passing scores that reflect the standard of minimum competency, NCEES conducts passing score studies on a regular basis. At these studies, a representative panel of engineers familiar with the candidate population uses a criterion-referenced procedure to set the passing score for the examination. The panel discusses the concept of minimum competence and develops a written standard that clearly articulates what skills and knowledge are required of engineers if they are to practice in a manner that protects the public health, safety, and welfare.

NCEES does not use fixed-percentage passing scores such as 70% or 75% because these fail to take into account the difficulty levels of different questions that make up the examinations. Similarly, NCEES avoids "grading on the curve" because licensure is designed to ensure that practitioners possess enough knowledge to perform professional activities in a manner that protects the public welfare. The key issue is whether an individual candidate is competent to practice and not whether the candidate is better or worse than other candidates.

The passing score can vary from one administration of the examination to another to reflect differences in difficulty levels of the examinations. However, the passing score is always based on the standard of minimum competency. To avoid confusion that might arise from fluctuations in the passing score, scores are converted to a standard scale which adopts 70 as the passing score. This technique of converting to a standard scale is commonly employed by testing specialists.

EXAMINATION PROCEDURES AND INSTRUCTIONS

Visit the NCEES Web site for current information about exam procedures and instructions.

Examination Materials

Before the morning and afternoon sessions, proctors will distribute examination booklets containing an answer sheet. You should not open the examination booklet until you are instructed to do so by the proctor. Read the instructions and information given on the front and back covers, and listen carefully to all the instructions the proctor reads.

The answer sheets for the multiple-choice questions are machine scored. For proper scoring, the answer spaces should be blackened completely. Since April 2002, NCEES has provided mechanical pencils with 0.7-mm HB lead to be used in the examination. You are not permitted to use any other writing instrument. If you decide to change an answer, you must erase the first answer completely. Incomplete erasures and stray marks may be read as intended answers. One side of the answer sheet is used to collect identification and biographical data. Proctors will guide you through the process of completing this portion of the answer sheet prior to taking the test. This process will take approximately 15 minutes.

Starting and Completing the Examination

You are not to open the examination booklet until instructed to do so by your proctor. If you complete the examination with more than 15 minutes remaining, you are free to leave after returning all examination materials to the proctor. Within 15 minutes of the end of the examination, you are required to rcmain until the end to avoid disruption to those still working and to permit orderly collection of all examination materials. Regardless of when you complete the examination, you are responsible for returning the numbered examination booklet assigned to you. Cooperate with the proctors collecting the examination materials. Nobody will be allowed to leave until the proctor has verified that all materials have been collected.

Calculators

Beginning with the April 2004 exam administration, the NCEES has strictly prohibited certain calculators from exam sites. Devices having a QWERTY keypad arrangement similar to a typewriter or keyboard are not permitted. Devices not permitted include but are not limited to palmtop, laptop, handheld, and desktop computers, calculators, databanks, data collectors, and organizers. The NCEES Web site (www.ncees.org) gives specific details on calculators.

Special Accommodations

The NCEES document *Guidelines for Requesting Religious and ADA Accommodations* explains the requirements for taking an NCEES exam with special testing accommodations. Candidates who wish to request special testing accommodations should refer to the NCEES Web site (www.ncees.org) under the "Exams" heading to find this document, along with frequently asked questions and forms for making the requests. To allow adequate evaluation time, NCEES must receive requests no later than 60 days prior to the exam administration.

EXAM SPECIFICATIONS FOR THE MORNING SESSION

NATIONAL COUNCIL OF EXAMINERS FOR ENGINEERING AND SURVEYING

Fundamentals of Engineering (FE) Examination

Effective October 2005

- The FE examination is an 8-hour supplied-reference examination: 120 questions in the 4-hour morning session and 60 questions in the 4-hour afternoon session.
- The afternoon session is administered in the following seven modules—Chemical, Civil, Electrical, Environmental, Industrial, Mechanical, and Other/General engineering.
- Examinees work all questions in the morning session and all questions in the afternoon module they have chosen.

MORNING SESSION
(120 QUESTIONS IN 12 TOPIC AREAS)

Topic Area	Approximate Percentage of Test Content
I. Mathematics	**15%**
A. Analytic geometry	
B. Integral calculus	
C. Matrix operations	
D. Roots of equations	
E. Vector analysis	
F. Differential equations	
G. Differential calculus	
II. Engineering Probability and Statistics	**7%**
A. Measures of central tendencies and dispersions (e.g., mean, mode, standard deviation)	
B. Probability distributions (e.g., discrete, continuous, normal, binomial)	
C. Conditional probabilities	
D. Estimation (e.g., point, confidence intervals) for a single mean	
E. Regression and curve fitting	
F. Expected value (weighted average) in decision-making	
G. Hypothesis testing	
III. Chemistry	**9%**
A. Nomenclature	
B. Oxidation and reduction	
C. Periodic table	
D. States of matter	
E. Acids and bases	
F. Equations (e.g., stoichiometry)	
G. Equilibrium	
H. Metals and nonmetals	

IV. Computers **7%**

A. Terminology (e.g., memory types, CPU, baud rates, Internet)
B. Spreadsheets (e.g., addresses, interpretation, "what if," copying formulas)
C. Structured programming (e.g., assignment statements, loops and branches, function calls)

V. Ethics and Business Practices **7%**

A. Code of ethics (professional and technical societies)
B. Agreements and contracts
C. Ethical versus legal
D. Professional liability
E. Public protection issues (e.g., licensing boards)

VI. Engineering Economics **8%**

A. Discounted cash flow (e.g., equivalence, PW, equivalent annual FW, rate of return)
B. Cost (e.g., incremental, average, sunk, estimating)
C. Analyses (e.g., breakeven, benefit-cost)
D. Uncertainty (e.g., expected value and risk)

VII. Engineering Mechanics (Statics and Dynamics) **10%**

A. Resultants of force systems
B. Centroid of area
C. Concurrent force systems
D. Equilibrium of rigid bodies
E. Frames and trusses
F. Area moments of inertia
G. Linear motion (e.g., force, mass, acceleration, momentum)
H. Angular motion (e.g., torque, inertia, acceleration, momentum)
I. Friction
J. Mass moments of inertia
K. Impulse and momentum applied to:
   1. particles
   2. rigid bodies
L. Work, energy, and power as applied to:
   1. particles
   2. rigid bodies

VIII. Strength of Materials **7%**

A. Shear and moment diagrams
B. Stress types (e.g., normal, shear, bending, torsion)
C. Stress strain caused by:
   1. axial loads
   2. bending loads
   3. torsion
   4. shear

D. Deformations (e.g., axial, bending, torsion)
E. Combined stresses
F. Columns
G. Indeterminant analysis
H. Plastic versus elastic deformation

IX. Material Properties **7%**

A. Properties
   1. chemical
   2. electrical
   3. mechanical
   4. physical
B. Corrosion mechanisms and control
C. Materials
   1. engineered materials
   2. ferrous metals
   3. nonferrous metals

X. Fluid Mechanics **7%**

A. Flow measurement
B. Fluid properties
C. Fluid statics
D. Energy, impulse, and momentum equations
E. Pipe and other internal flow

XI. Electricity and Magnetism **9%**

A. Charge, energy, current, voltage, power
B. Work done in moving a charge in an electric field (relationship between voltage and work)
C. Force between charges
D. Current and voltage laws (Kirchhoff, Ohm)
E. Equivalent circuits (series, parallel)
F. Capacitance and inductance
G. Reactance and impedance, susceptance and admittance
H. AC circuits
I. Basic complex algebra

XII. Thermodynamics **7%**

A. Thermodynamic laws (e.g., 1st Law, 2nd Law)
B. Energy, heat, and work
C. Availability and reversibility
D. Cycles
E. Ideal gases
F. Mixture of gases
G. Phase changes
H. Heat transfer
I. Properties of:
   1. enthalpy
   2. entropy

MORNING SAMPLE QUESTIONS

NOTE: THESE QUESTIONS REPRESENT ONLY HALF THE NUMBER OF QUESTIONS THAT APPEAR ON THE ACTUAL EXAMINATION.

MORNING SAMPLE QUESTIONS

1. If the functional form of a curve is known, differentiation can be used to determine all of the following **EXCEPT** the:

(A) concavity of the curve

(B) location of inflection points on the curve

(C) number of inflection points on the curve

(D) area under the curve between certain bounds

2. Which of the following is the general solution to the differential equation and boundary condition shown below?

$$\frac{dy}{dt} + 5y = 0;\ y(0) = 1$$

(A) e^{5t}

(B) e^{-5t}

(C) $e^{\sqrt{-5t}}$

(D) $5e^{-5t}$

3. If D is the differential operator, then the general solution to $(D + 2)^2 y = 0$ is:

(A) $C_1 e^{-4x}$

(B) $C_1 e^{-2x}$

(C) $e^{-4x}(C_1 + C_2 x)$

(D) $e^{-2x}(C_1 + C_2 x)$

4. A particle traveled in a straight line in such a way that its distance S from a given point on that line after time t was $S = 20t^3 - t^4$. The rate of change of acceleration at time $t = 2$ is:

(A) 72
(B) 144
(C) 192
(D) 208

5. Which of the following is a unit vector perpendicular to the plane determined by the vectors $\mathbf{A} = 2\mathbf{i} + 4\mathbf{j}$ and $\mathbf{B} = \mathbf{i} + \mathbf{j} - \mathbf{k}$?

(A) $-2\mathbf{i} + \mathbf{j} - \mathbf{k}$

(B) $\frac{1}{\sqrt{5}}(\mathbf{i} + 2\mathbf{j})$

(C) $\frac{1}{\sqrt{6}}\left(-2\mathbf{i} + \mathbf{j} - \mathbf{k}\right)$

(D) $\frac{1}{\sqrt{6}}\left(-2\mathbf{i} - \mathbf{j} - \mathbf{k}\right)$

6. If f' denotes the derivative of a function of $y = f(x)$, then $f'(x)$ is defined by:

(A) $\lim\limits_{\Delta y \to 0} \frac{\Delta x}{\Delta y}$

(B) $\lim\limits_{\Delta y \to 0} \frac{\Delta y}{\Delta x}$

(C) $\lim\limits_{\Delta x \to 0} \frac{f(x + \Delta x) - f(x)}{\Delta x}$

(D) $\lim\limits_{\Delta y \to 0} \frac{f(x - \Delta x) + f(x)}{\Delta y}$

GO ON TO THE NEXT PAGE

7. What is the area of the region in the first quadrant that is bounded by the line $y = 1$, the curve $x = y^{3/2}$, and the y-axis?

(A) 2/5
(B) 3/5
(C) 2/3
(D) 1

8. Three lines are defined by the three equations:

$$\begin{aligned} x + y &= 0 \\ x - y &= 0 \\ 2x + y &= 1 \end{aligned}$$

The three lines form a triangle with vertices at:

(A) $(0, 0), \left(\frac{1}{3}, \frac{1}{3}\right), (1, -1)$

(B) $(0, 0), \left(\frac{2}{3}, \frac{2}{3}\right), (-1, -1)$

(C) $(1, 1), (1, -1), (2, 1)$

(D) $(1, 1), (3, -3), (-2, -1)$

GO ON TO THE NEXT PAGE

9. The value of the integral $\int_0^{\pi} 10 \sin x dx$ is:

(A) -10
(B) 0
(C) 10
(D) 20

10. You wish to estimate the mean M of a population from a sample of size n drawn from the population. For the sample, the mean is x and the standard deviation is s. The probable accuracy of the estimate improves with an increase in:

(A) M
(B) n
(C) s
(D) $M + s$

11. A bag contains 100 balls numbered from 1 to 100. One ball is removed. What is the probability that the number on this ball is odd or greater than 80?

(A) 0.2
(B) 0.5
(C) 0.6
(D) 0.8

12. The standard deviation of the population of the three values 1, 4, and 7 is:

(A) $\sqrt{3}$

(B) $\sqrt{6}$

(C) 4

(D) 6

13. Suppose the lengths of telephone calls form a normal distribution with a mean length of 8.0 min and a standard deviation of 2.5 min. The probability that a telephone call selected at random will last more than 15.5 min is most nearly:

(A) 0.0013
(B) 0.0026
(C) 0.2600
(D) 0.9987

14. The volume (L) of 1 mol of H_2O at 546 K and 1.00 atm pressure is most nearly:

(A) 11.2
(B) 14.9
(C) 22.4
(D) 44.8

GO ON TO THE NEXT PAGE

15. Consider the equation:

$$As_2O_3 + 3\ C \rightarrow 3\ CO + 2\ As$$

Atomic weights may be taken as 75 for arsenic, 16 for oxygen, and 12 for carbon. According to the equation above, the reaction of 1 standard gram-mole of As_2O_3 with carbon will result in the formation of:

(A) 1 gram-mole of As

(B) 28 grams of CO

(C) 150 grams of As

(D) a greater amount by weight of CO than of As

16. If 60 mL of NaOH solution neutralizes 40 mL of 0.50 M H_2SO_4, the concentration of the NaOH solution is most nearly:

(A) 0.80 M
(B) 0.67 M
(C) 0.45 M
(D) 0.33 M

GO ON TO THE NEXT PAGE

17. The atomic weights of sodium, oxygen, and hydrogen are 23, 16 and 1, respectively. To neutralize 4 grams of NaOH dissolved in 1 L of water requires 1 L of:

(A) 0.001 normal HCl solution

(B) 0.01 normal HCl solution

(C) 0.1 normal HCl solution

(D) 1.0 normal HCl solution

18. Consider the following equation:

$$K = \frac{[C]^2[D]^2}{[A]^4[B]}$$

The equation above is the formulation of the chemical equilibrium constant equation for which of the following reactions?

(A) $C_2 + D_2 \leftrightarrow A_4 + B$

(B) $4A + B \leftrightarrow 2C + 2D$

(C) $4C + 2D \leftrightarrow 2A + B$

(D) $A_4 + B \leftrightarrow C_2 + D_2$

GO ON TO THE NEXT PAGE

19. The flowchart for a computer program contains the following segment:

```
   VAR = 0
┌→ IF VAR < 5 THEN VAR = VAR + 2
│  OTHERWISE EXIT LOOP
└─ LOOP
```

What is the value of VAR at the conclusion of this routine?

(A) 0
(B) 2
(C) 4
(D) 6

20. In a spreadsheet, the number in Cell A4 is set to 6. Then A5 is set to A4 + A4. This formula is copied into Cells A6 and A7. The number shown in Cell A7 is most nearly:

(A) 12
(B) 24
(C) 36
(D) 216

21. Consider the following program segment:

```
INPUT Z, N
S = 1
T = 1
FOR K = 1 TO N
T = T*Z/K
S = S + T
NEXT K
```

This segment calculates the sum:

(A) $S = 1 + ZT + 2\ ZT + 3\ ZT + \ldots + N\ ZT$

(B) $S = 1 + ZT + \frac{1}{2}ZT + \frac{1}{3}ZT + \ldots + \left(\frac{1}{N}\right)ZT$

(C) $S = 1 + \frac{Z}{1} + \frac{2Z}{2} + \frac{3Z}{3} + \ldots + \left(\frac{NZ}{N}\right)$

(D) $S = 1 + \frac{Z}{1!} + \frac{Z^2}{2!} + \frac{Z^3}{3!} + \ldots + \left(\frac{Z^N}{N!}\right)$

22. In a spreadsheet, Row 1 has the numbers 2, 4, 6, 8, 10, ... , 20 in Columns A–J, and Row 2 has the numbers 1, 3, 5, 7, 9, ... , 19 in the same columns. All other cells are zero except for Cell D3, which contains the formula: D1 + D$1*D2. This formula has been copied into cells D4 and D5. The number that appears in cell D4 is most nearly:

(A) 3
(B) 64
(C) 519
(D) 4,216

GO ON TO THE NEXT PAGE

23. An engineer testifying as an expert witness in a product liability case should:

(A) answer as briefly as possible only those questions posed by the attorneys

(B) provide a complete and objective analysis within his or her area of competence

(C) provide an evaluation of the character of the defendant

(D) provide information on the professional background of the defendant

24. A professional engineer, originally licensed 30 years ago, is asked to evaluate a newly developed computerized control system for a public transportation system. The engineer may accept this project if:

(A) he or she is competent in the area of modern control systems

(B) his or her professional engineering license has not lapsed

(C) his or her original area of specialization was in transportation systems

(D) he or she has regularly attended meetings of a professional engineering society

GO ON TO THE NEXT PAGE

25. You and your design group are competing for a multidisciplinary concept project. Your firm is the lead group in the design professional consortium formed to compete for the project. Your consortium has been selected as the first to enter fee negotiations with the project owner. During the negotiations, the amount you have to cut from your fee to be awarded the contract will require dropping one of the consortium members whose staff has special capabilities not available from the staff of the remaining consortium members. Can your remaining consortium ethically accept the contract?

(A) No, because an engineer may not accept a contract to coordinate a project with other professional firms providing capabilities and services that must be provided by hired consultants.

(B) Yes, if your remaining consortium members hire a few new lower-cost employees to do the special work that would have been provided by the consortium member that has been dropped.

(C) No, not if the owner is left with the impression that the consortium is still fully qualified to perform all the required tasks.

(D) Yes, if in accepting an assignment to coordinate the project, a single person will sign and seal all the documents for the entire work of the consortium.

26. You have an on-site job interview to follow up on an on-campus interview with Company A. Just before you fly to the interview, you get a call from Company B asking you to come for an on-site interview at their offices in the same city. When you inform them of your interview with Company A, they suggest you stop in after that. Company A has already paid for your airfare and, at the conclusion of your interview with them, issues you reimbursement forms for the balance of your trip expenses with instructions to file for all your trip expenses. When you inform them of your added interview stop at Company B, they tell you to go ahead and charge the entire cost of the trip to Company A. You interview with Company B, and at the conclusion, they give you travel reimbursement forms with instructions to file for all your trip expenses. When you inform them of the instructions of Company A, they tell you that the only expenses requiring receipts are airfare and hotel rooms, so you should still file for all the other expenses with them even if Company A is paying for it because students always need a little spending money. What should you do?

(A) Try to divide the expenses between both firms as best you can.

(B) Do as both recruiting officers told you. It is their money and their travel policies.

(C) File for travel expenses with only one firm.

(D) Tell all your classmates to sign up to interview with these firms for the trips.

27. A company can manufacture a product using hand tools. Costs will be $1,000 for tools and a $1.50 manufacturing cost per unit. As an alternative, an automated system will cost $15,000 with a $0.50 manufacturing cost per unit. With an anticipated annual volume of 5,000 units and neglecting interest, the breakeven point (years) is most nearly:

(A) 2.8
(B) 3.6
(C) 15.0
(D) never

28. A printer costs $900. Its salvage value after 5 years is $300. Annual maintenance is $50. If the interest rate is 8%, the equivalent uniform annual cost is most nearly:

(A) $224
(B) $300
(C) $327
(D) $350

29. The need for a large-capacity water supply system is forecast to occur 4 years from now. At that time, the system required is estimated to cost $40,000. If an account earns 12% per year compounded annually, the amount that must be placed in the account at the end of each year in order to accumulate the necessary purchase price is most nearly:

(A) $8,000
(B) $8,370
(C) $9,000
(D) $10,000

GO ON TO THE NEXT PAGE

30. A project has the estimated cash flows shown below.

Year End	0	1	2	3	4
Cash Flow	–$1,100	–$400	+$1,000	+$1,000	+$1,000

Using an interest rate of 12% per year compounded annually, the annual worth of the project is most nearly:

(A) $450
(B) $361
(C) $320
(D) $226

31. You must choose between four pieces of comparable equipment based on the cash flows given below. All four pieces have a life of 8 years.

Parameter	**Equipment**			
	A	**B**	**C**	**D**
First cost	$25,000	$35,000	$20,000	$40,000
Annual costs	$ 8,000	$ 6,000	$ 9,000	$ 5,000
Salvage value	$ 2,500	$ 3,500	$ 2,000	$ 4,000

The discount rate is 12%. Ignore taxes. The most preferable top two projects and the difference between their present worth values are most nearly:

(A) A and C, $170
(B) B and D, $170
(C) A and C, $234
(D) B and D, $234

32. Referring to the figure below, the coefficient of static friction between the block and the inclined plane is 0.25. The block is in equilibrium.

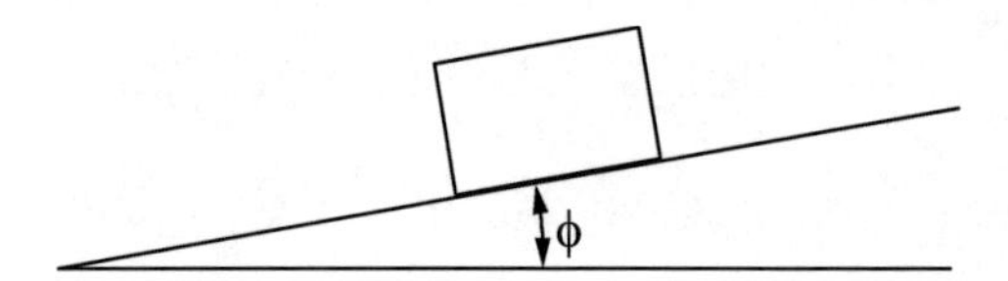

As the inclined plane is raised, the block will begin to slide when:

(A) $\sin \phi = 1.0$

(B) $\cos \phi = 1.0$

(C) $\cos \phi = 0.25$

(D) $\tan \phi = 0.25$

33. A cylinder weighing 120 N rests between two frictionless walls as shown in the following figure.

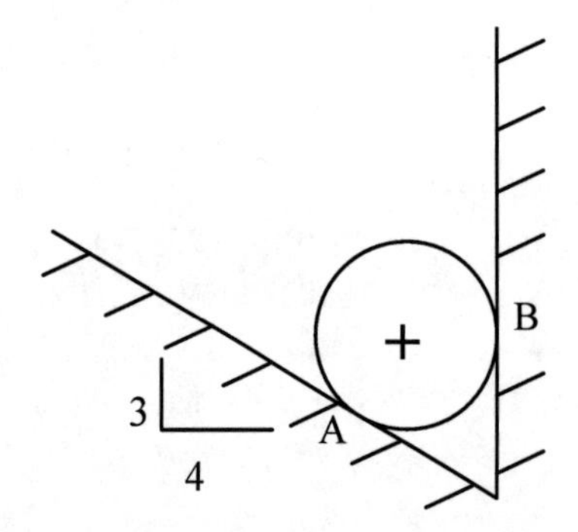

The wall reaction (N) at Point A is most nearly:

(A) 96
(B) 139
(C) 150
(D) 200

GO ON TO THE NEXT PAGE

34. Three forces act as shown below.

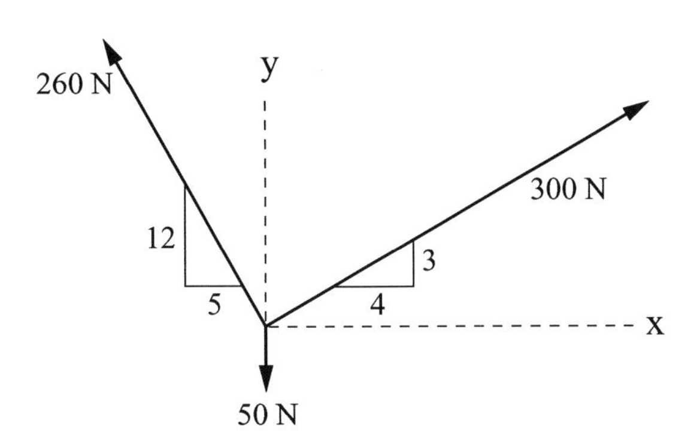

The magnitude of the resultant of the three forces (N) is most nearly:

(A) 140
(B) 191
(C) 370
(D) 396

35. In the figure below, Block A weighs 50 N, Block B weighs 80 N, and Block C weighs 100 N. The coefficient of friction at all surfaces is 0.30.

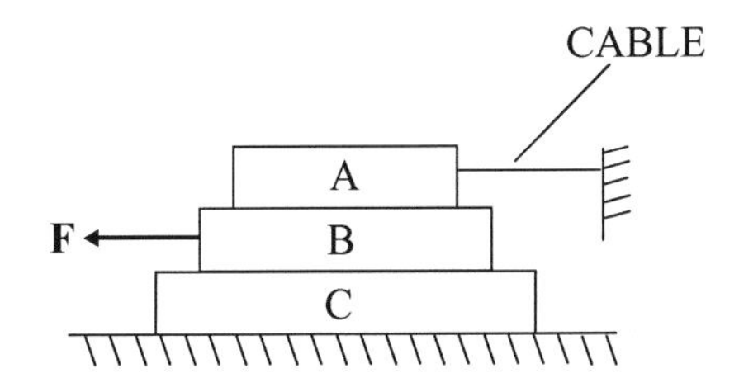

The maximum force **F** (N) that can be applied to Block B without disturbing equilibrium is most nearly:

(A) 15
(B) 54
(C) 69
(D) 84

36. The moment of force **F** (N•m) shown below with respect to Point p is most nearly:

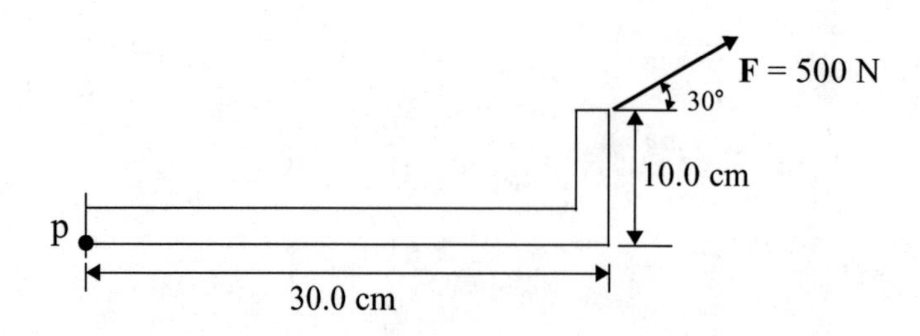

(A) 31.7 ccw
(B) 31.7 cw
(C) 43.3 cw
(D) 43.3 ccw

37. The figure below shows a simple truss.

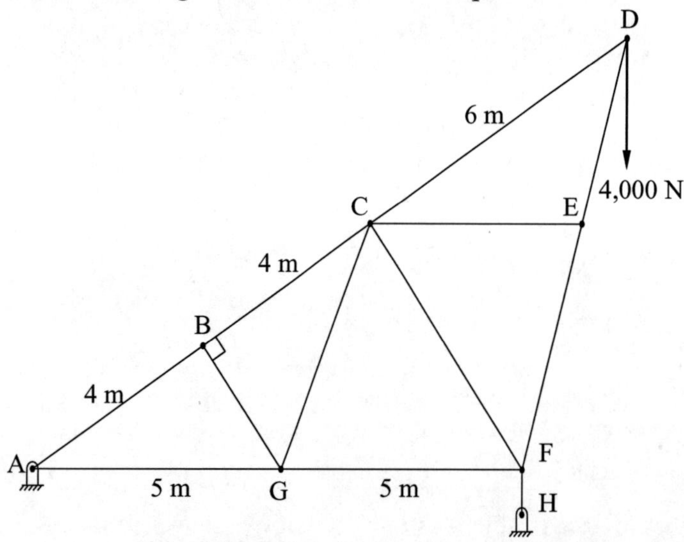

The zero-force members in the truss are:

(A) BG, CG, CF, CE
(B) BG, CE
(C) CF
(D) CG, CF

GO ON TO THE NEXT PAGE

38. The beam shown below is known as a:

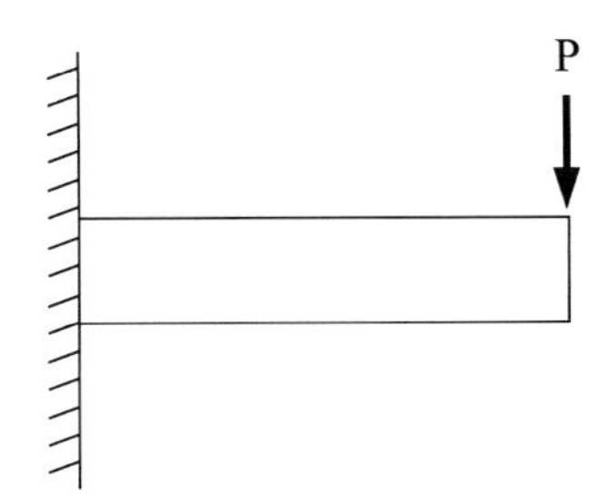

(A) cantilever beam
(B) statically indeterminate beam
(C) simply supported beam
(D) continuously loaded beam

39. The shear diagram for a particular beam is shown below. All lines in the diagram are straight. The bending moment at each end of the beam is zero, and there are no concentrated couples along the beam.

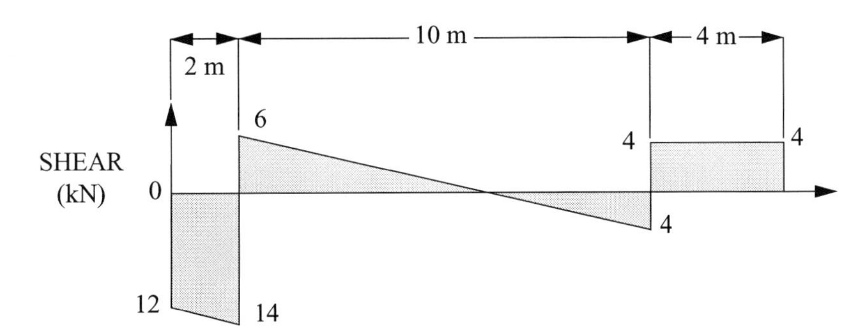

The maximum magnitude of the bending moment (kN•m) in the beam is most nearly:

(A) 8
(B) 16
(C) 18
(D) 26

40. The piston of a steam engine is 50 cm in diameter, and the maximum steam gage pressure is 1.4 MPa. If the design stress for the piston rod is 68 MPa, its cross-sectional area (m^2) should be most nearly:

(A) 40.4×10^{-4}
(B) 98.8×10^{-4}
(C) 228.0×10^{-4}
(D) 323.0×10^{-4}

41. A shaft of wood is to be used in a certain process. If the allowable shearing stress parallel to the grain of the wood is 840 kN/m^2, the torque (N•m) transmitted by a 200-mm-diameter shaft with the grain parallel to the neutral axis is most nearly:

(A) 500
(B) 1,200
(C) 1,320
(D) 1,500

42. The Euler formula for columns deals with:

(A) relatively short columns

(B) shear stress

(C) tensile stress

(D) elastic buckling

43. The mechanical deformation of a material above its recrystallization temperature is commonly known as:

(A) hot working
(B) strain aging
(C) grain growth
(D) cold working

44. In general, a metal with high hardness will also have:

(A) good formability
(B) high impact strength
(C) high electrical conductivity
(D) high yield strength

45. Glass is said to be an amorphous material. This means that it:

(A) has a high melting point
(B) is a supercooled vapor
(C) has large cubic crystals
(D) has no apparent crystal structure

GO ON TO THE NEXT PAGE

46. If an aluminum crimp connector were used to connect a copper wire to a battery, what would you expect to happen?

(A) The copper wire only will corrode.

(B) The aluminum connector only will corrode.

(C) Both will corrode.

(D) Nothing

47. The rectangular homogeneous gate shown below is 3.00 m high × 1.00 m wide and has a frictionless hinge at the bottom. If the fluid on the left side of the gate has a density of 1,600 kg/m^3, the magnitude of the force **F** (kN) required to keep the gate closed is most nearly:

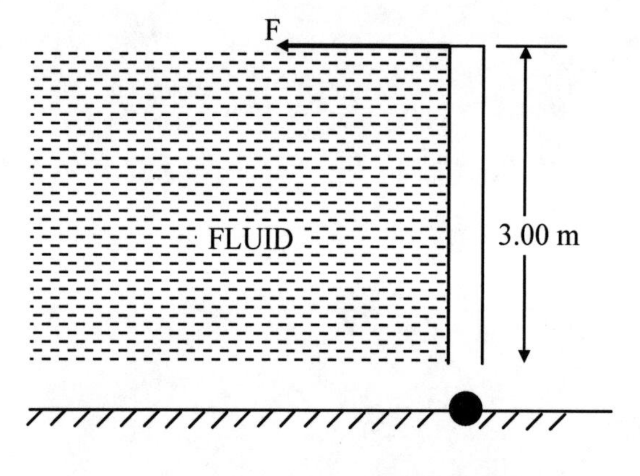

(A) 0
(B) 22
(C) 24
(D) 220

48. Which of the following statements is true of viscosity?

(A) It is the ratio of inertial to viscous force.

(B) It always has a large effect on the value of the friction factor.

(C) It is the ratio of the shear stress to the rate of shear deformation.

(D) It is usually low when turbulent forces predominate.

49. A horizontal jet of water (density = 1,000 kg/m^3) is deflected perpendicularly to the original jet stream by a plate as shown below.

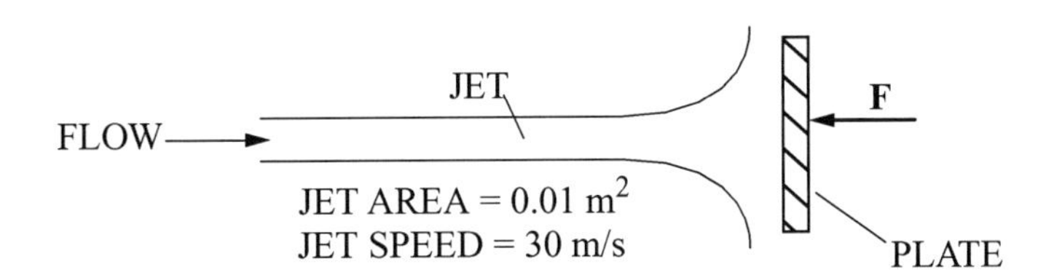

The magnitude of force **F** (kN) required to hold the plate in place is most nearly:

(A) 4.5
(B) 9.0
(C) 45.0
(D) 90.0

50. Which of the following statements about flow through an insulated valve is most accurate?

(A) The enthalpy rises.

(B) The upstream and downstream enthalpies are equal.

(C) Temperature increases sharply.

(D) Pressure increases sharply.

51. The pitot tube shown below is placed at a point where the velocity is 2.0 m/s. The specific gravity of the fluid is 2.0, and the upper portion of the manometer contains air. The reading h (m) on the manometer is most nearly:

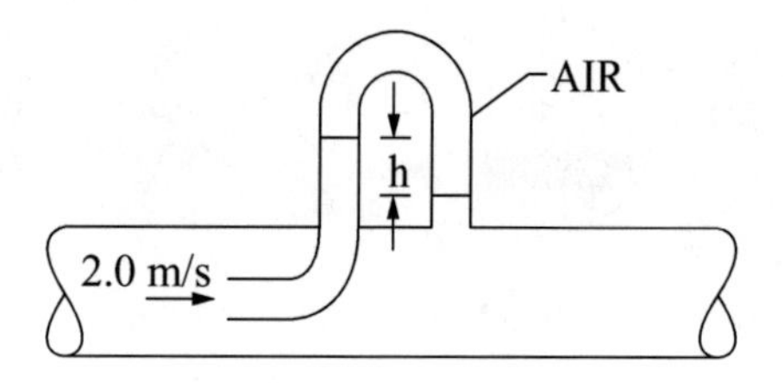

(A) 20.0
(B) 10.0
(C) 0.40
(D) 0.20

52. If the complex power is 1,500 VA with a power factor of 0.866 lagging, the reactive power (VAR) is most nearly:

(A) 0
(B) 750
(C) 1,300
(D) 1,500

53. Series-connected circuit elements are shown in the figure below.

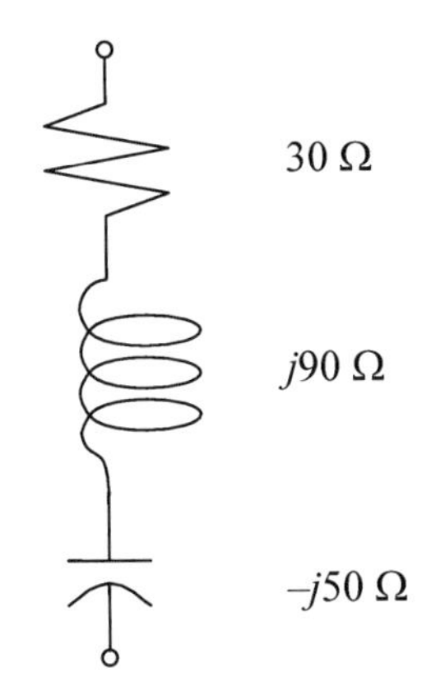

Which of the following impedance diagrams is correct according to conventional notation?

(A)

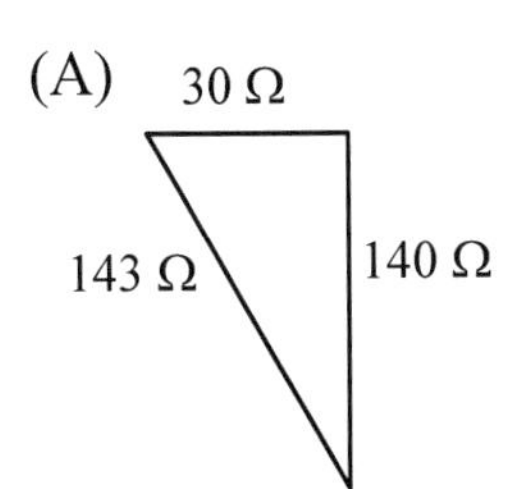

(B)

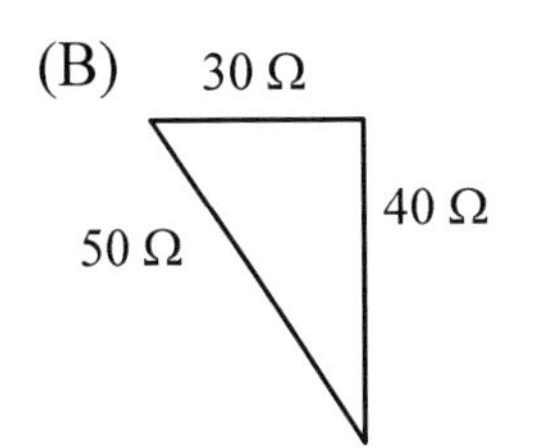

(C)

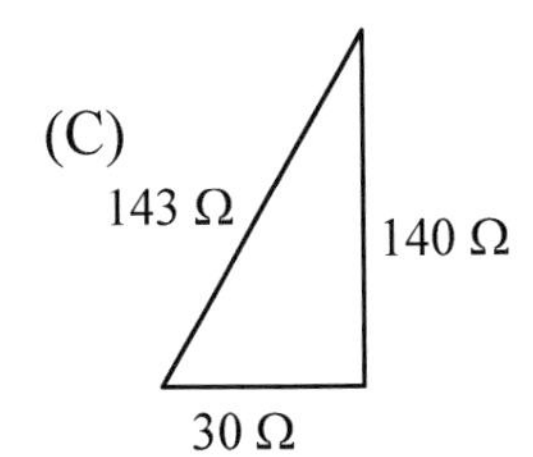

(D)

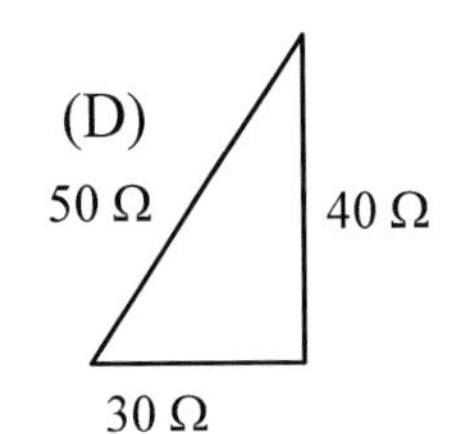

54. A 10-μF capacitor has been charged to a potential of 150 V. A resistor of 25 Ω is then connected across the capacitor through a switch. When the switch is closed for ten time constants, the total energy (joules) dissipated by the resistor is most nearly:

(A) 1.0×10^{-7}
(B) 1.1×10^{-1}
(C) 9.0×10^{1}
(D) 9.0×10^{3}

55. The connecting wires and the battery in the circuit shown below have negligible resistance.

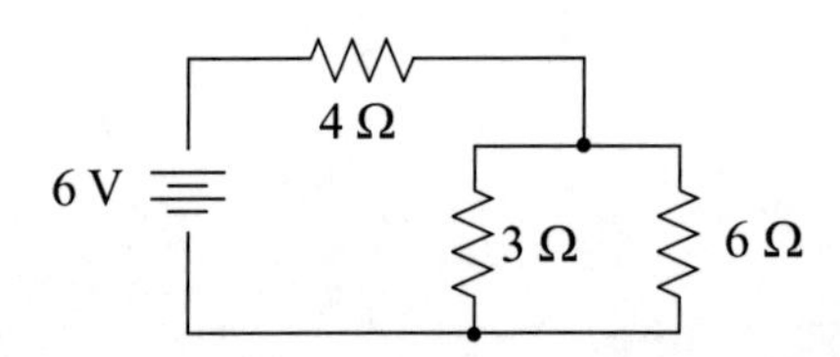

The current (amperes) through the 6-Ω resistor is most nearly:

(A) 1/3
(B) 1/2
(C) 1
(D) 3/2

56. The term $\frac{(1-i)^2}{(1+i)^2}$, where $i=\sqrt{-1}$, is most nearly:

(A) -1
(B) $-1 + i$
(C) 0
(D) $1 + i$

57. An insulated tank contains half liquid and half vapor by volume in equilibrium. The release of a small quantity of the vapor without the addition of heat will cause:

(A) evaporation of some liquid in the tank
(B) superheating of the vapor in the tank
(C) a rise in temperature
(D) an increase in enthalpy

58. The heat transfer during an adiabatic process is:

(A) reversible
(B) irreversible
(C) dependent on temperature
(D) zero

59. An isentropic process is one which:

(A) is adiabatic but not reversible

(B) is reversible but not adiabatic

(C) is adiabatic and reversible

(D) occurs at constant pressure and temperature

60. The universal gas constant is 8.314 kJ/(kmol•K). The gas constant [(kJ/(kg•K)] of a gas having a molecular weight of 44 is most nearly:

(A) 0.19
(B) 0.38
(C) 0.55
(D) 5.3

IF YOU FINISH BEFORE TIME IS CALLED, YOU MAY WISH TO CHECK YOUR WORK ON THIS TEST.

MORNING SOLUTIONS

ANSWERS TO THE MORNING QUESTIONS

Detailed solutions for each question begin on the next page.

QUESTION	ANSWER	QUESTION	ANSWER
1	D	31	B
2	B	32	D
3	D	33	C
4	A	34	D
5	C	35	B
6	C	36	A
7	A	37	A
8	A	38	A
9	D	39	D
10	B	40	A
11	C	41	C
12	B	42	D
13	A	43	A
14	D	44	D
15	C	45	D
16	B	46	B
17	C	47	C
18	B	48	C
19	D	49	B
20	B	50	B
21	D	51	D
22	C	52	B
23	B	53	D
24	A	54	B
25	A	55	A
26	A	56	A
27	A	57	A
28	A	58	D
29	B	59	C
30	D	60	A

MORNING SOLUTIONS

1. The area under a curve is determined by integration, not differentiation.

THE CORRECT ANSWER IS: (D)

2. The characteristic equation for a first-order linear homogeneous differential equation is:

$$r + 5 = 0$$

which has a root at $r = -5$.

Refer to Differential Equations in the Mathematics section of the *FE Reference Handbook*. The form of the solution is then:

$$y = C\,e^{-\alpha t} \text{ where } \alpha = a \quad \text{and} \quad a = 5 \text{ for this problem}$$

C is determined from the boundary condition.

$$1 = C\,e^{-5(0)}$$
$$C = 1$$

Then, $y = e^{-5t}$

THE CORRECT ANSWER IS: (B)

3. Refer to Differential Equations in the Mathematics section of the *FE Reference Handbook*. The characteristic equation for a second-order linear homogeneous differential equation is:

$$r^2 + 2ar + b = 0$$

In this problem, $D^2 + 4D + 4 = 0$, so:

$$2a = 4 \text{ or } a = 2 \text{ and } b = 4$$

In solving the characteristic equation, it is noted that there are repeated real roots: $r_1 = r_2 = -2$

Because $a^2 = b$, the solution for this critically damped system is:

$$y(x) = (C_1 + C_2\,x)\,e^{-2x}$$

THE CORRECT ANSWER IS: (D)

MORNING SOLUTIONS

4. First, the velocity is:

$$V = S' = 60t^2 - 4t^3$$

Then, the acceleration is:

$$A = S'' = 120t - 12t^2$$

Finally, the rate of change of acceleration is:

$$A' = S''' = 120 - 24t$$

When $t = 2$:

$$A' = 120 - 24(2) = 120 - 48 = 72$$

THE CORRECT ANSWER IS: (A)

5. The cross product of vectors **A** and **B** is a vector perpendicular to **A** and **B**.

$$\begin{vmatrix} \mathbf{i} & \mathbf{j} & \mathbf{k} \\ 2 & 4 & 0 \\ 1 & 1 & -1 \end{vmatrix} = \mathbf{i}(-4) - \mathbf{j}(-2-0) + \mathbf{k}(2-4) = -4\mathbf{i} + 2\mathbf{j} - 2\mathbf{k}$$

To obtain a unit vector, divide by the magnitude.

$$\text{Magnitude} = \sqrt{(-4)^2 + 2^2 + (-2)^2} = \sqrt{24} = 2\sqrt{6}$$

$$\frac{-4\mathbf{i} + 2\mathbf{j} - 2\mathbf{k}}{2\sqrt{6}} = \frac{-2\mathbf{i} + \mathbf{j} - \mathbf{k}}{\sqrt{6}}$$

THE CORRECT ANSWER IS: (C)

6. Refer to Differential Calculus in the Mathematics section of the *FE Reference Handbook.*

THE CORRECT ANSWER IS: (C)

MORNING SOLUTIONS

7. Define a differential strip with length $(x - 0)$ and height dy.

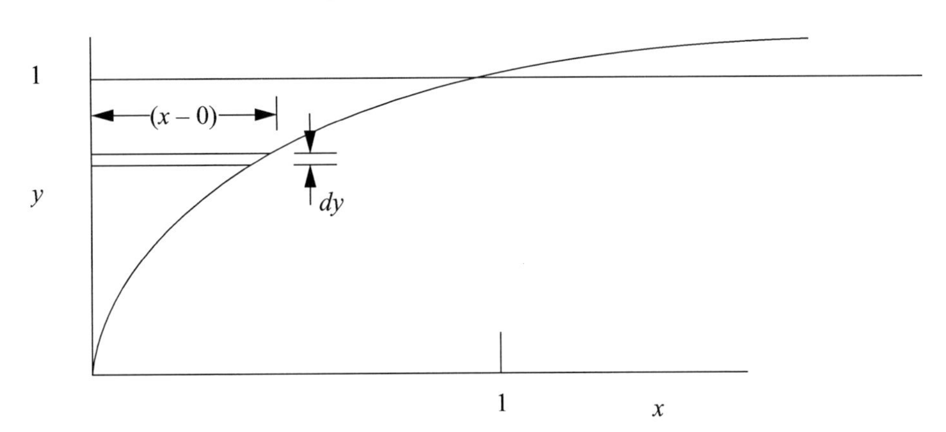

$$\int dA = \int_0^1 x\,dy = \int_0^1 y^{3/2}\,dy = \left.\frac{y^{5/2}}{5/2}\right|_0^1 = \frac{2}{5}$$

THE CORRECT ANSWER IS: (A)

8.

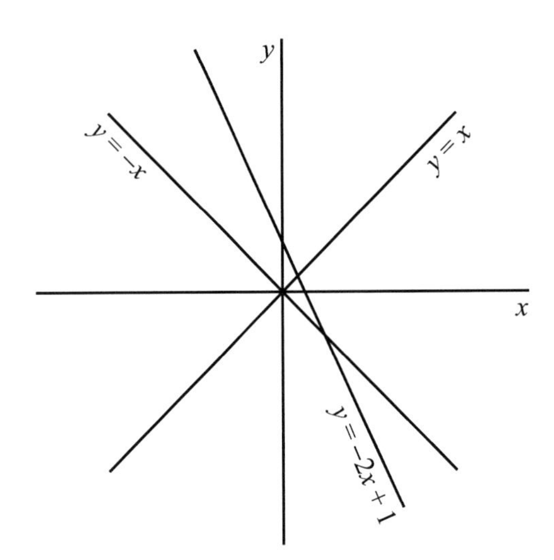

$y = -x$

$y = x$

$y = -2x + 1$

from graph one, intersection is at (0,0), so (C) and (D) are incorrect.

Also, second intersection is at (1,–1), so key has to be (A).

(0,0) (1/3, 1/3) and (1,–1)

THE CORRECT ANSWER IS: (A)

9. $\int_0^{\pi} 10 \sin x dx = 10\left[-\cos x \Big|_0^{\pi}\right]$

$$= 10\left[-\cos \pi - (-\cos 0)\right]$$

$$= 10[1+1]$$

$$= 20$$

THE CORRECT ANSWER IS: (D)

10. Accuracy increases with increasing sample size.

THE CORRECT ANSWER IS: (B)

11. The key word is **OR**. What is the probability that the number is odd **OR** greater than 80? Refer to Property 2 given under Probability and Statistics in the Mathematics section of the *FE Reference Handbook.*

P(A + B) = P(A) + P(B) – P(A,B)

Event A is removing a ball with an odd number.
P(A) = 50/100 = 0.5

Event B is removing a ball with a number greater than 80.
P(B) = 20/100 = 0.2

Event A,B is removing a ball with an odd number that is greater than 80.

There are ten such balls.
P(A,B) = 10/100 = 0.1

Also P(A,B) = P(A) × P(B) = 0.5 × 0.2 = 0.1

P(A + B) = 0.5 + 0.2 – (0.5 × 0.2) = 0.6

THE CORRECT ANSWER IS: (C)

12.

x	$x-\bar{x}$	$(x-\bar{x})^2$
1	–3	9
4	0	0
7	3	9
$\sum = 12$		$\sum = 18$

$$\bar{X} = \frac{12}{3} = 4$$

$$\sigma = \sqrt{\frac{18}{3}} = \sqrt{6}$$

THE CORRECT ANSWER IS: (B)

13. $8 - 15.5 = 7.5$

$$\frac{7.5}{2.5} = 3 \text{ standard deviations}$$

From the Unit Normal Distribution Table in the Mathematics section of the *FE Reference Handbook*.

For X = 3, R(X) = 0.0013

THE CORRECT ANSWER IS: (A)

14. PV = nRT

1V = (1)(0.08206)(546)

solve for V = 44.8 L

THE CORRECT ANSWER IS: (D)

15. 2 moles of As × 75 g/mole of As = 150 g of As

THE CORRECT ANSWER IS: (C)

16. $H_2SO_4 + 2\ NaOH \rightarrow Na_2SO_4 + 2\ H_2O$
$0.5\ M\ H_2SO_4 = 1.0\ N\ H_2SO_4$

$1.0\ M\ NaOH = 1.0\ N\ NaOH$

$40\ mL$ of $1.0\ N\ H_2SO_4 = 60\ mL$ of $x\ N\ NaOH$

$40 \times 1 = 60x$

$x = 40/60 = 0.67\ N = 0.67\ M\ NaOH$

THE CORRECT ANSWER IS: (B)

17. The molecular weight of NaOH is 40 g; therefore, 4 g/L of NaOH will form 1 L of 0.1 normal NaOH solution. One liter of 0.1 normal HCl solution is required to neutralize the NaOH.

THE CORRECT ANSWER IS: (C)

18. Refer to the Chemistry section of the *FE Reference Handbook* for the equilibrium constant of a chemical reaction.

$4A + B \leftrightarrow 2C + 2D$

THE CORRECT ANSWER IS: (B)

19.

Step	VAR
1	0
2	2
3	4
4	6

EXIT LOOP

At the conclusion of the routine, VAR = 6.

THE CORRECT ANSWER IS: (D)

20.

Row	Column A	Value of A
4	6	6
5	A4 + A4	12
6	A5 + A4	18
7	A6 + A4	24

THE CORRECT ANSWER IS: (B)

21.

Step	Z	N	T	K	S
1	Z	N	.	.	.
2	Z	N	1	.	1
3	Z	N	1	1	1
.	Z	N	Z	1	1
.			Z	1	1 + Z
(NEXT K)			$\frac{Z^2}{2}$	2	$\frac{1+Z+Z^2}{2}$
(NEXT K)			$\frac{Z^3}{(2)(3)}$	3	$\frac{\frac{1+Z+Z^2}{2+Z^3}}{(2)(3)}$
(NEXT K)			$\frac{Z^4}{(2)(3)(4)}$	4	$\frac{\frac{\frac{1+Z+Z^2}{2+Z^3}}{(2)(3)+Z^3}}{(2)(3)(4)}$

$\therefore$ The sequence is: $S = 1 + \frac{Z}{1!} + \frac{Z^2}{2!} + \frac{Z^3}{3!} + \frac{Z^4}{4!} + \ldots + \frac{Z^N}{N!}$

THE CORRECT ANSWER IS: (D)

22.

Rows	Columns				
	A	B	C	D	E
1	2	4	6	8	10
2	1	3	5	7	9
3				64	
4				519	
5					

D3: D1 + D$1 * D2 = 8 + 8(7) = 64

D4: D2 + D$1 * D3 = 7 + 8(64) = 519

THE CORRECT ANSWER IS: (C)

23. Refer to the NCEES *Model Rules of Professional Conduct*, Section A.4. in the Ethics section of the *FE Reference Handbook.*

THE CORRECT ANSWER IS: (B)

24. Refer to the NCEES *Model Rules of Professional Conduct*, Section B.1. in the Ethics section of the *FE Reference Handbook.*

THE CORRECT ANSWER IS: (A)

25. Refer to the NCEES *Model Rules of Professional Conduct*, Section B.3. and Section C.1. in the Ethics section of the *FE Reference Handbook.*

THE CORRECT ANSWER IS: (A)

26. Refer to the NCEES *Model Rules of Professional Conduct*, Section B.5. and Section B.6. in the Ethics section of the *FE Reference Handbook.*

THE CORRECT ANSWER IS: (A)

27. \$1.50 (5,000) = \$7,500

\$0.50 (5,000) = \$2,500

Annual savings = \$7,500 – \$2,500 = \$5,000

Additional investment = \$15,000 – \$1,000 = \$14,000

Payback = \$14,000/\$5,000 = 2.8 years

THE CORRECT ANSWER IS: (A)

28. Annual cost: = \$900(A/P, 8%, 5) + \$50 – \$300(A/F, 8%, 5)
= \$900(0.2505) + \$50 – \$300(0.1705)
= \$225.45 + \$50 – \$51.15
= \$224.30

THE CORRECT ANSWER IS: (A)

29.

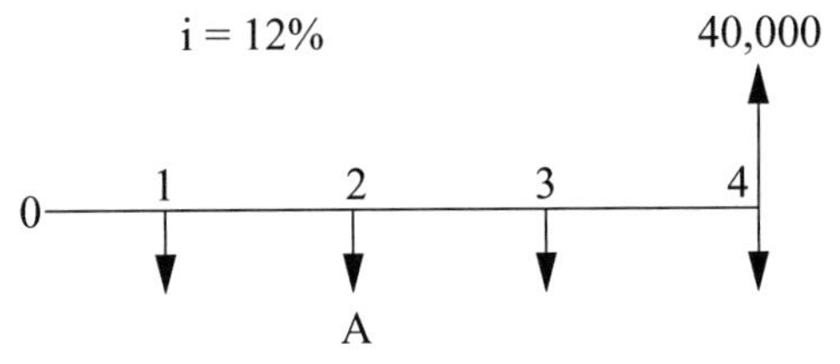

i = 12%
40,000
0 1 2 3 4
A

$A = F(A/F, i, n) = 40,000(A/F, 12\%, 4) = \$8,369$ per year

THE CORRECT ANSWER IS: (B)

30.

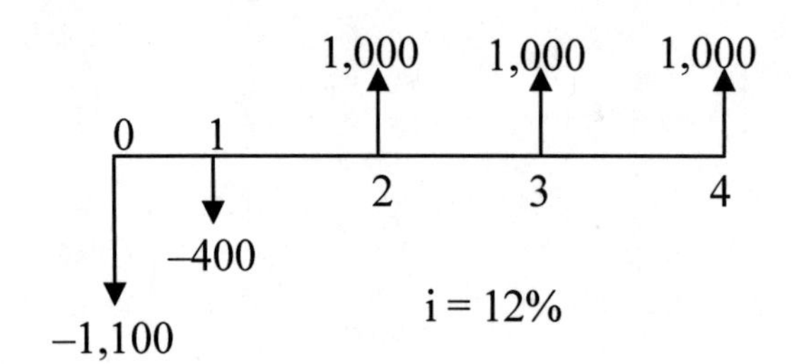

$$PW = -1{,}100 - 400\,(P/F, 12\%, 1) + 1{,}000\,(P/F, 12\%, 2)$$

$$+\ 1{,}000\,(P/F, 12\%, 3) + 1{,}000\,(P/F, 12\%, 4)$$

$$= -1{,}100 - 400\,(0.8929) + 1{,}000\,(0.7972) + 1{,}000\,(0.7118) + 1{,}000\,(0.6355)$$

$$= 687.34$$

$$A = PW\,(A/P, 12\%, 4) = 687.34(0.3292)$$

$$= \$226 \text{ per year}$$

THE CORRECT ANSWER IS: (D)

31. The easiest way to solve this problem is to look at the present worth of each alternative.

The present worth values are all given by

PW = First Cost + Annual Cost × (P/A, 12%, 8) – Salvage Value × (P/F, 12%, 8)
= First Cost + Annual Cost × 4.9676 – Salvage Value × 0.4039

Then PW(A) = $63,731
PW(B) = $63,392
PW(C) = $63,901
PW(D) = $63,222

The cash flows are all costs, so the most preferable two projects, those with the lowest present worth costs, are B and D, and the difference between them is $170.

THE CORRECT ANSWER IS: (B)

MORNING SOLUTIONS

32. Normal to the plane:

$$\Sigma F_n = 0: N - mg \cos \phi = 0 \rightarrow N = mg \cos \phi$$

Tangent to the plane:

$$\Sigma F_t = 0: -mg \sin \phi + \mu N = 0$$

$$\therefore -mg \sin \phi + \mu mg \cos \phi = 0$$

$$\frac{\sin \phi}{\cos \phi} = \tan \phi = \mu$$

$$\tan \phi = 0.25$$

THE CORRECT ANSWER IS: (D)

33. $\Sigma F_y = 0 = -120 + \frac{4}{5} A$

$A = 150$ N

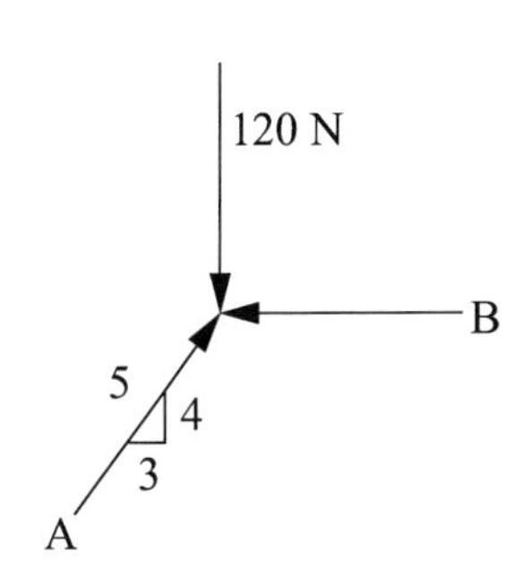

THE CORRECT ANSWER IS: (C)

34. $R_y = \Sigma F_y = \frac{12}{13}(260) + \frac{3}{5}(300) - 50 = 370$

$R_x = \Sigma F_x = -\frac{5}{13}(260) + \frac{4}{5}(300) = 140$

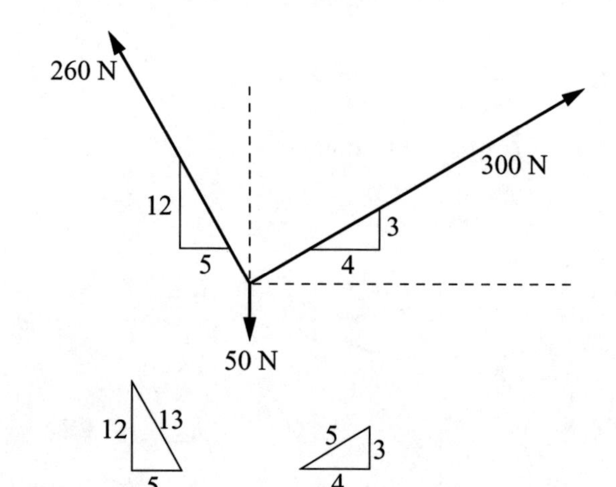

$R = \sqrt{R_x^2 + R_y^2} = \sqrt{370^2 + 140^2}$

$R = 396 \text{ N}$

THE CORRECT ANSWER IS: (D)

35.

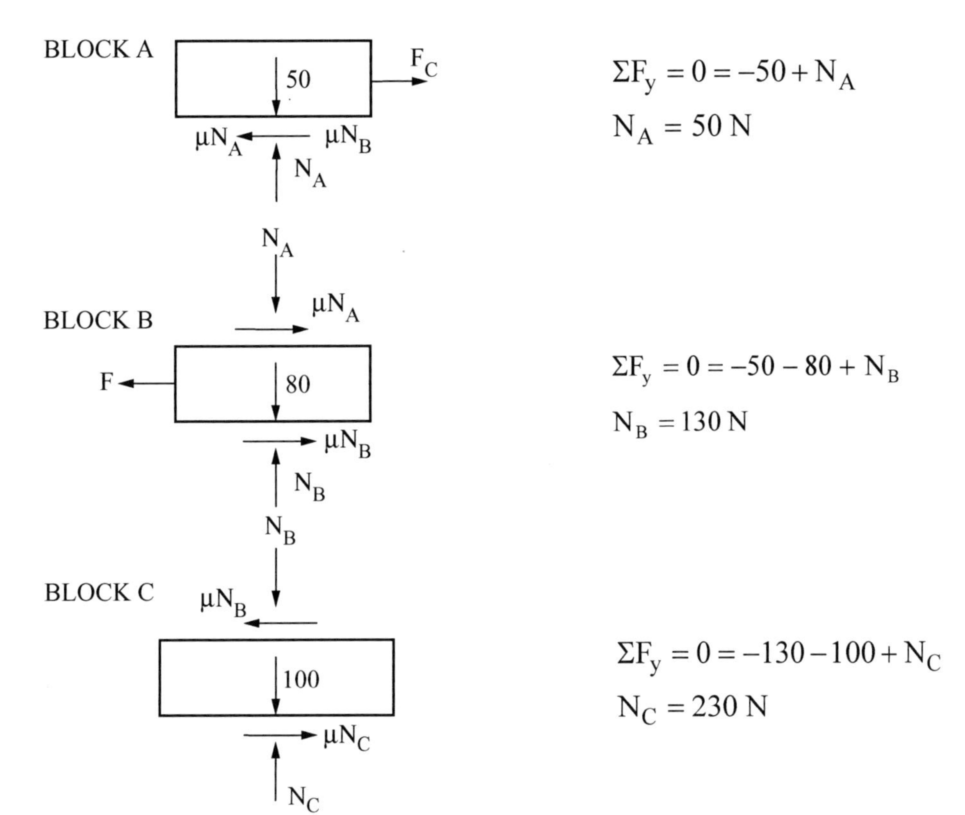

$$\Sigma F_y = 0 = -50 + N_A$$

$$N_A = 50 \text{ N}$$

$$\Sigma F_y = 0 = -50 - 80 + N_B$$

$$N_B = 130 \text{ N}$$

$$\Sigma F_y = 0 = -130 - 100 + N_C$$

$$N_C = 230 \text{ N}$$

Assume Blocks A and C remain stationary.

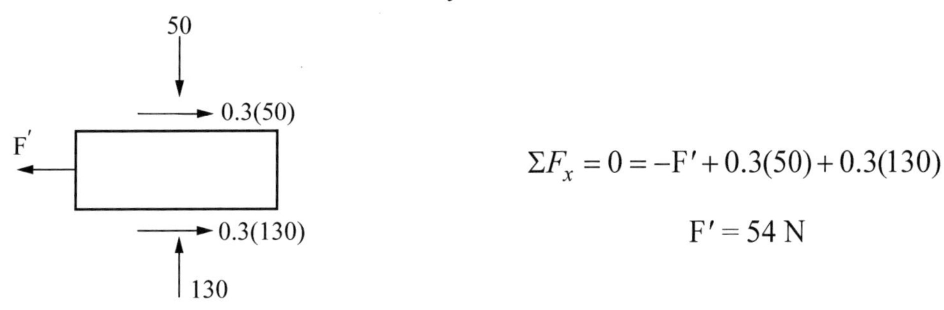

$$\Sigma F_x = 0 = -F' + 0.3(50) + 0.3(130)$$

$$F' = 54 \text{ N}$$

Assume Blocks B and C move.

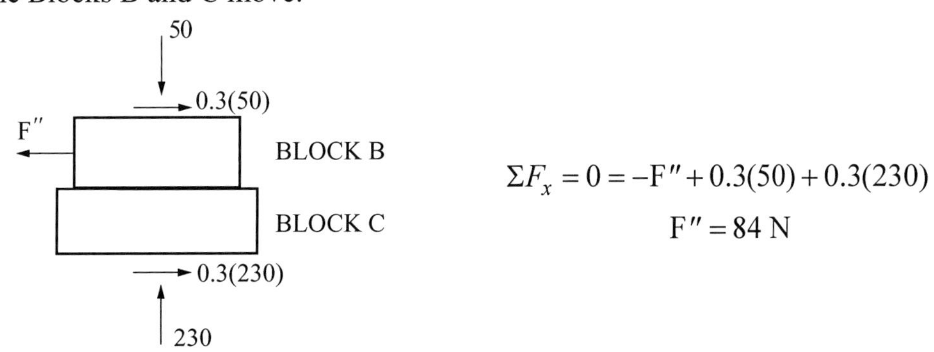

$$\Sigma F_x = 0 = -F'' + 0.3(50) + 0.3(230)$$

$$F'' = 84 \text{ N}$$

$\therefore$ **F** = 54 N where A and C remain stationary.

THE CORRECT ANSWER IS: (B)

Morning Solutions

36. $F_H = 500 \cos 30° = 433$

$F_V = 500 \sin 30° = 250$

$M_P = 250(0.30) - 433(0.10) = 31.7$ N•m ccw

THE CORRECT ANSWER IS: (A)

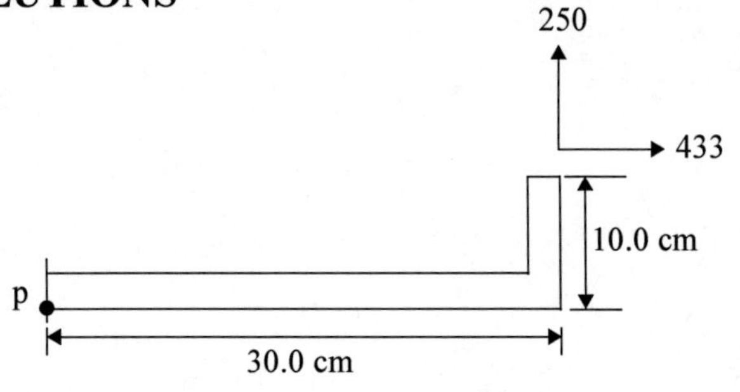

37. Zero-force members usually occur at joints where members are aligned as follows:

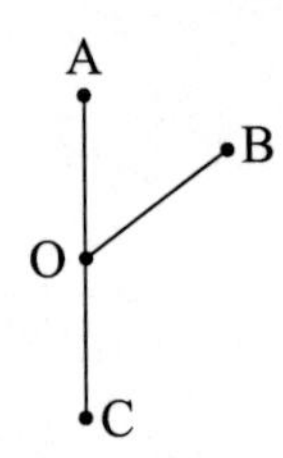

That is, joints where two members are along the same line (OA and OC) and the third member is at some arbitrary angle. That member (OB) is a zero-force member because the forces in OA and OC must be equal and opposite.

For this specific problem, we immediately examine joints B and E:

B: 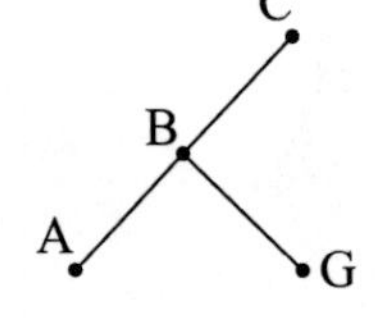

BG is zero-force member

E: 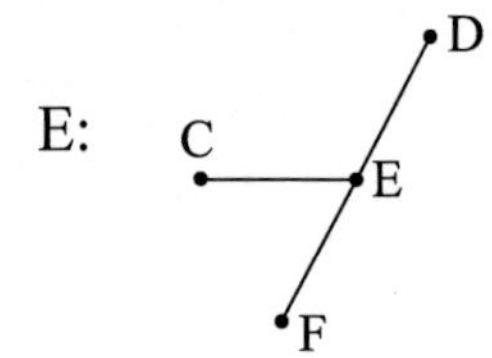

CE is zero-force member

Now, examine Joint G. Since BG is zero-force member, the joint effectively looks like:

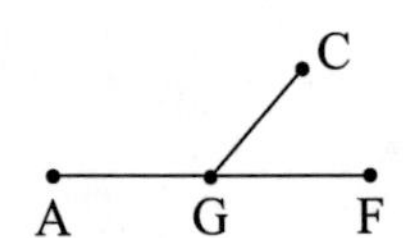

and, therefore, CG is another zero-force member.

Finally, examine Joint C. Since both CG and CE are zero force members, the joint effectively looks like:

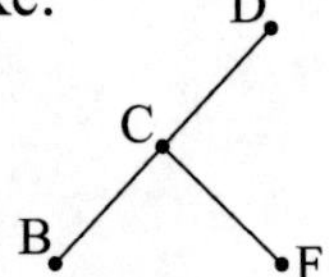

and, therefore, CF is another zero-force member. Thus, BG, CE, CG, CF are the zero-force members.

THE CORRECT ANSWER IS: (A)

MORNING SOLUTIONS

38. By definition of a cantilever beam, it is NOT statically indeterminate, it is completely supported, and it is loaded only at a specific point.

THE CORRECT ANSWER IS: (A)

39. $$\frac{10 \text{ m}}{10 \text{ kN}} = \frac{x}{6 \text{ kN}}$$

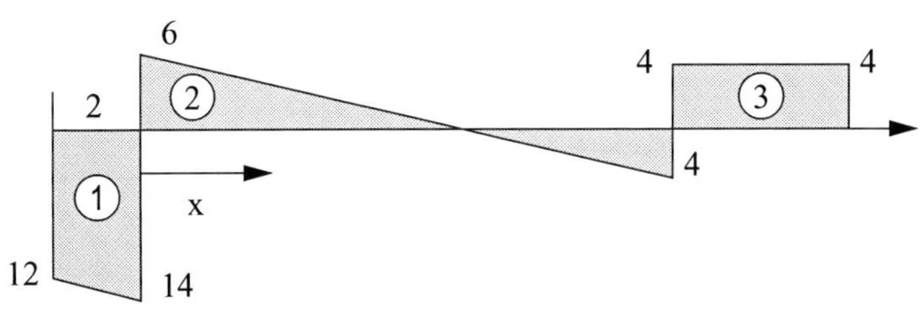

$x = 6 \text{ m}$

$\text{Area } 1 = 13(2) = 26 \text{ kN•m}$

$$\text{Area } 2 = \frac{6(6)}{2} = 18 \text{ kN•m}$$

$\text{Area } 3 = 4(4) = 16 \text{ kN•m}$

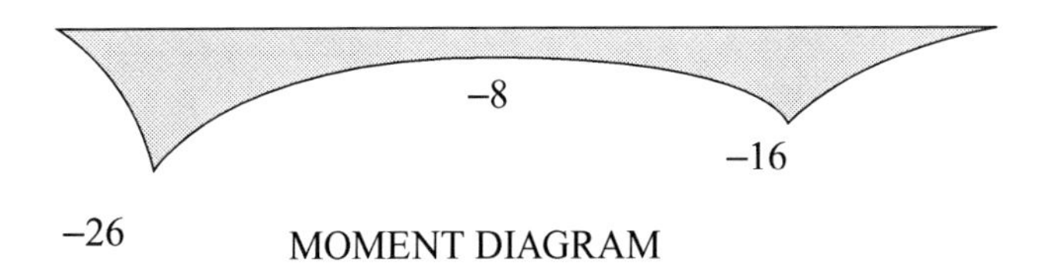

Maximum magnitude of the bending moment is 26 kN•m.

THE CORRECT ANSWER IS: (D)

40. $$\Sigma F = PA = \left(1.4 \times 10^6\right)\left(\frac{\pi(0.5)^2}{4}\right) = F_{rod}$$

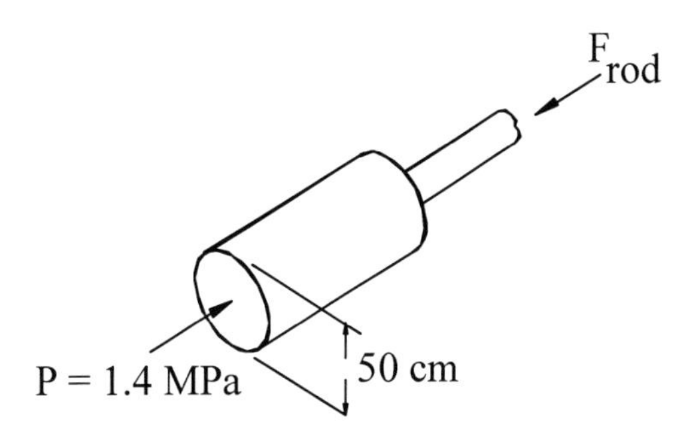

$F_{rod} = 275 \text{ kN} = \sigma A = 68 \times 10^6 A$

$A = 40.4 \times 10^{-4} \text{m}^2$

THE CORRECT ANSWER IS: (A)

MORNING SOLUTIONS

41. $$\tau = \frac{Tr}{J} = \frac{T\frac{d}{2}}{\frac{\pi d^4}{32}} = \frac{16T}{\pi d^3}$$

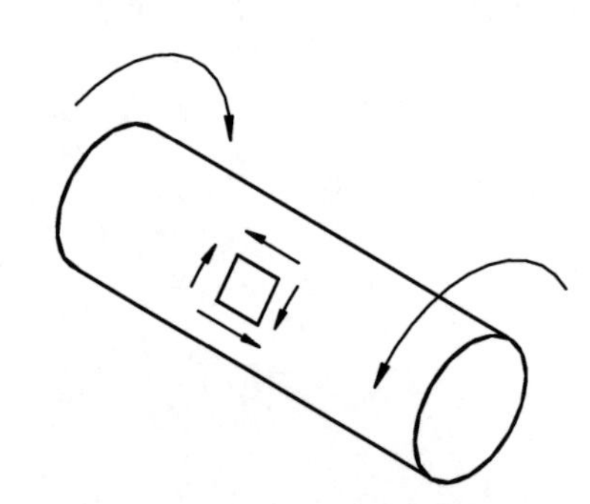

$$T = \frac{\pi d^3 \tau}{16} = \frac{\pi (0.2)^3 (840 \times 10^3)}{16}$$

$$T = 1{,}319 \text{ N}\bullet\text{m}$$

THE CORRECT ANSWER IS: (C)

42. The Euler formula is used for elastic stability of relatively long columns, subjected to concentric axial loads in compression.

THE CORRECT ANSWER IS: (D)

43. The question statement is the definition of hot working.

THE CORRECT ANSWER IS: (A)

44. By definition, a metal with high hardness has a high tensile strength and a high yield strength.

THE CORRECT ANSWER IS: (D)

45. By definition, amorphous materials do not have a crystal structure.

THE CORRECT ANSWER IS: (D)

46. Aluminum is anodic relative to copper and, therefore, will corrode to protect the copper.

THE CORRECT ANSWER IS: (B)

MORNING SOLUTIONS

47. The mean pressure of the fluid acting on the gate is evaluated at the mean height, and the center of pressure is 2/3 of the height from the top; thus, the total force of the fluid is:

$$F_f = \rho g \frac{H}{2}(H) = 1,600(9.807)\frac{3}{2}(3) = 70,610 \text{ N}$$

and its point of application is 1.00 m above the hinge. A moment balance about the hinge gives:

$$F(3) - F_f(1) = 0$$

$$F = \frac{F_f}{3} = \frac{70,610}{3} = 23,537 \text{ N}$$

THE CORRECT ANSWER IS: (C)

48. Refer to the Fluid Mechanics section of the *FE Reference Handbook.*

$$\tau_t = \mu\left(\frac{dv}{dy}\right)$$

where τ_t = shear stress and

$\frac{dv}{dy}$ = rate of shear deformation

Hence, μ is the ratio of shear stress to the rate of shear deformation.

THE CORRECT ANSWER IS: (C)

49. $Q = A_1V_1 = (0.01\ m^2)(30\ m/s)$

$= 0.3\ m^3/s$

Since the water jet is deflected perpendicularly, the force F must deflect the total horizontal momentum of the water.

$F = \rho QV = (1{,}000\ kg/m^3)\ (0.3\ m^3/s)\ (30\ m/s) = 9{,}000\ N = 9.0\ kN$

THE CORRECT ANSWER IS: (B)

50. Flow through an insulated valve closely follows a throttling process. A throttling process is at constant enthalpy.

THE CORRECT ANSWER IS: (B)

51. $\frac{\rho v^2}{2} = gh\left(\rho - \rho_{air}\right)$

$$\therefore h = \frac{\rho v^2}{2g\left(\rho - \rho_{air}\right)} \approx \frac{v^2}{2g} \approx \frac{(2)^2}{(2)(9.8)} \approx 0.204\ m$$

THE CORRECT ANSWER IS: (D)

52. S = apparent power
P = real power
Q = reactive power

$S = P + jQ = |S|\cos\theta + j\,|S|\sin\theta$

$\cos\theta = pf = 0.866$

$Q = (1{,}500\ VA)\ \sin[\cos^{-1}0.866] = 750\ VAR$

THE CORRECT ANSWER IS: (B)

MORNING SOLUTIONS

53. $Z = 30 + j90 - j50 = 30 + j40\ \Omega$

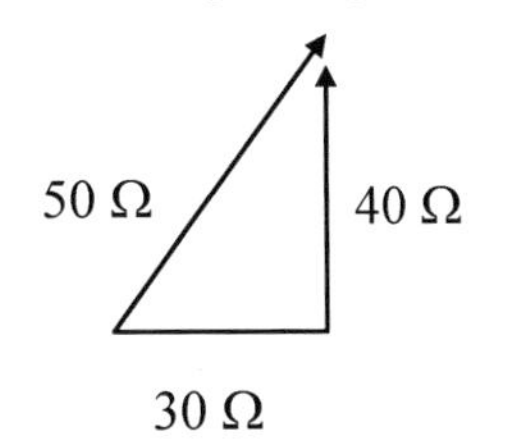

THE CORRECT ANSWER IS: (D)

54. Initially, $V_C(t) = 150$ V

$$W_C(t) = \frac{1}{2}cV_C^2 = \frac{1}{2}(10 \times 10^{-6} F)(150\ V)^2$$

$W_C = 0.113$ J initial stored energy.

After ten time constants, all energy will be dissipated.

THE CORRECT ANSWER IS: (B)

55. $R_T = 4\Omega + 3\Omega \,||\, 6\Omega = 4\Omega + 2\Omega$

$$R_T = 6\Omega \Rightarrow I_T = \frac{6\ V}{6\Omega} = 1\ A$$

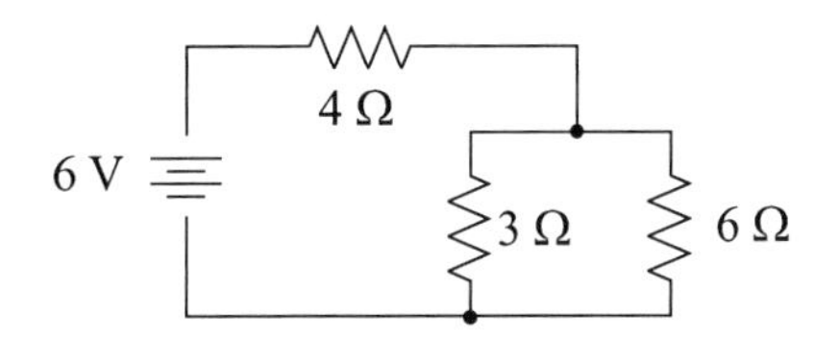

$$I_x = \frac{3}{9}(I_T) = \frac{1}{3}A$$

THE CORRECT ANSWER IS: (A)

56. $$\frac{(1-i)^2}{(1+i)^2} = \frac{1-2i+i^2}{1+2i+i^2} = \frac{1-1-2i}{1-1+2i} = \frac{-i}{i} = -1$$

THE CORRECT ANSWER IS: (A)

MORNING SOLUTIONS

57. As vapor escapes, the mass within the tank is reduced. With constant volume, the specific volume within the tank must increase. This can happen only if liquid evaporates.

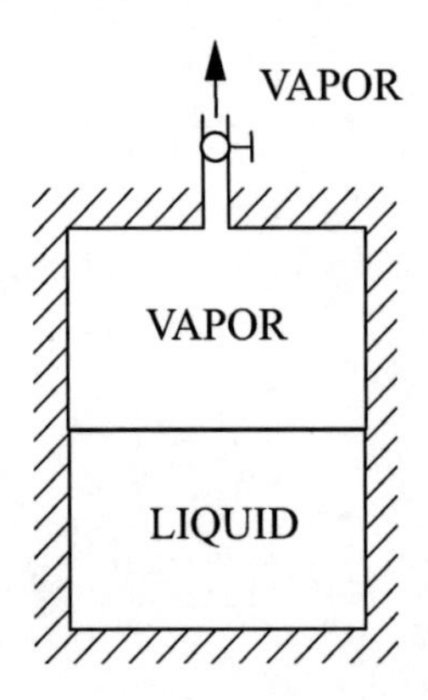

THE CORRECT ANSWER IS: (A)

58. By definition, an adiabatic process is a process in which no heat is transferred.

THE CORRECT ANSWER IS: (D)

59. An isentropic process is one for which the entropy remains constant. Entropy is defined by the equation:

$$dS = \left(\frac{\delta Q}{T}\right)_{\text{reversible}}$$

The entropy will be constant if $\delta Q = 0$ and the process is reversible. It is theoretically possible for a nonadiabatic, irreversible process to have a constant entropy, but this is not one of the responses. Option (D) describes a state, not a process.

THE CORRECT ANSWER IS: (C)

60. $$R = \frac{Ru}{M} = \frac{8.314}{44} = 0.1890 \frac{\text{kJ}}{\text{kg}\bullet\text{K}}$$

THE CORRECT ANSWER IS: (A)

EXAM SPECIFICATIONS FOR THE AFTERNOON SESSION

AFTERNOON SESSION IN INDUSTRIAL ENGINEERING
(60 questions in 8 topic areas)

Topic Area	Approximate Percentage of Test Content
I. Engineering Economics	**15%**
II. Probability and Statistics	**15%**
III. Modeling and Computation	**12%**

I. Engineering Economics — **15%**

A. Discounted cash flows (equivalence, PW, EAC, FW, IRR, loan amortization)
B. Types and breakdown of costs (e.g., fixed, variable, direct and indirect labor, material, capitalized)
C. Analyses (e.g., benefit-cost, breakeven, minimum cost, overhead, risk, incremental, life cycle)
D. Accounting (financial statements and overhead cost allocation)
E. Cost estimating
F. Depreciation and taxes
G. Capital budgeting

II. Probability and Statistics — **15%**

A. Combinatorics (e.g., combinations, permutations)
B. Probability distributions (e.g., normal, binomial, empirical)
C. Conditional probabilities
D. Sampling distributions, sample sizes, and statistics (e.g., central tendency, dispersion)
E. Estimation (point estimates, confidence intervals)
F. Hypothesis testing
G. Regression (linear, multiple)
H. System reliability (single components, parallel and series systems)
I. Design of experiments (e.g., ANOVA, factorial designs)

III. Modeling and Computation — **12%**

A. Algorithm and logic development (e.g., flow charts, pseudo-code)
B. Spreadsheets
C. Databases (e.g., types, information content, relational)
D. Decision theory (e.g., uncertainty, risk, utility, decision trees)
E. Optimization modeling (decision variables, objective functions, and constraints)
F. Linear programming (e.g., formulation, primal, dual, graphical solution)
G. Math programming (network, integer, dynamic, transportation, assignment)
H. Stochastic models (e.g., queuing, Markov, reliability)
I. Simulation (e.g., event, process, Monte Carlo sampling, random number generation, steady-state vs. transient)

IV. Industrial Management — 10%

A. Principles (e.g., planning, organizing) and tools of management (e.g., MBO, re-engineering)
B. Organizational structure (e.g., functional, matrix, line/staff)
C. Motivation theories (e.g., Maslow, Theory X, Theory Y)
D. Job evaluation and compensation
E. Project management (scheduling, PERT, CPM)

V. Manufacturing and Production Systems — 13%

A. Manufacturing systems (e.g., cellular, group technology, flexible, lean)
B. Process design (e.g., number of machines/people, equipment selection, and line balancing)
C. Inventory analysis (e.g., EOQ, safety stock)
D. Forecasting
E. Scheduling (e.g., sequencing, cycle time, material control)
F. Aggregate planning (e.g., JIT, MRP, MRPII, ERP)
G. Concurrent engineering and design for manufacturing
H. Automation concepts (e.g., robotics, CIM)
I. Economics (e.g., profits and costs under various demand rates, machine selection)

VI. Facilities and Logistics — 12%

A. Flow measurements and analysis (e.g., from/to charts, flow planning)
B. Layouts (e.g., types, distance metrics, planning, evaluation)
C. Location analysis (e.g., single facility location, multiple facility location, storage location within a facility)
D. Process capacity analysis (e.g., number of machines/people, trade-offs)
E. Material handling capacity analysis (storage & transport)
F. Supply chain design (e.g., warehousing, transportation, inventories)

VII. Human Factors, Productivity, Ergonomics, and Work Design — 12%

A. Methods analysis (e.g., improvement, charting) and task analysis (e.g., MTM, MOST)
B. Time study (e.g., time standards, allowances)
C. Workstation design
D. Work sampling
E. Learning curves
F. Productivity measures
G. Risk factor identification, safety, toxicology, material safety data sheets (MSDS)
H. Environmental stress assessment (e.g., noise, vibrations, heat, computer-related)
I. Design of tasks, tools, displays, controls, user interfaces, etc.
J. Anthropometry, biomechanics, and lifting

VIII. Quality **11%**

A. Total quality management theory (e.g., Deming, Juran) and application
B. Management and planning tools (e.g., fishbone, Pareto, quality function deployment, scatter diagrams)
C. Control charts
D. Process capability and specifications
E. Sampling plans
F. Design of experiments for quality improvement
G. Auditing, ISO certification, and the Baldridge award

INDUSTRIAL
AFTERNOON SAMPLE QUESTIONS

NOTE: THESE QUESTIONS REPRESENT HALF THE NUMBER OF QUESTIONS THAT APPEAR ON THE ACTUAL EXAMINATION.

INDUSTRIAL SAMPLE QUESTIONS

1. Management is considering the purchase of a new automated loading machine for an assembly line. The following information on the new purchase from the local distributor of the machine is known:

I. The new machine will cost $25,000 and have a $5,000 salvage value at the end of 5 years.
II. The new machine will reduce the cost of labor by 25% but increase the cost of electricity by 40%.
III. The company's annual expenses for the loading operations are currently $45,000 for labor and $2,500 for electricity.

Assuming that the cost of capital is 10%, the net present value of the purchase decision is most nearly:

(A) –$19,191.50
(B) $16,960.20
(C) $21,908.20
(D) $52,498.90

2. You are helping a sheltered workshop find a profitable product, and plan to make wooden dump trucks. The average selling price of a finished truck is $15.00. Operations in making the truck include sawing (4 min), drilling (2 min), and painting (3 min), with one person able to perform each operation. Labor costs $10.00/hr including all overhead. The saw cost $1,200, the drill $300, and the paint booth $1,500. Materials costs are estimated as $1.50/truck. The minimum quantity of trucks needed to break even on these costs is most nearly:

(A) 125
(B) 200
(C) 222
(D) 250

3. The following variables are defined for a typical metal cutting operation:

M = machine and operator charge rate ($/hour)
N_T = number of tools used (number of tools/batch)
N_b = number of parts in a batch (number of parts/batch)
t_L = part loading/unloading time (hours/part)
t_m = machining time (hours/part)
t_c = tool change time for worn tools (hours/tool)
C_t = tool cost ($/tool)

The total production cost per part C_{pr} can best be described as:

(A) $Mt_L + Mt_m + M(N_T/N_b)t_c + (N_T/N_b)C_t$
(B) $M(N_bt_L + N_bt_c + N_Tt_m) + N_TC_t$
(C) $M(N_bt_L + N_bt_m + N_Tt_c) + N_TC_t$
(D) $Mt_L + Mt_m + (N_T/N_b)t_c + (N_T/N_b)C_t$

GO ON TO THE NEXT PAGE

4. The estimated costs for a bridge are as follows:

Initial construction: \$800,000
Renovation every five years: \$300,000
Annual upkeep in addition to renovation: \$75,000

At an annual interest rate of 8%, the capitalized cost of this bridge is most nearly:

(A) \$926,150
(B) \$1,576,875
(C) \$2,376,875
(D) \$2,487,500

5. A dimension under control is assumed to be normally distributed with a mean of 6.55 and a standard deviation of 0.15. The acceptable range for this dimension is 6.45–6.80. The percent of product that is nonconforming is most nearly:

(A) 4.7%
(B) 24.4%
(C) 25.2%
(D) 29.9%

GO ON TO THE NEXT PAGE

6. A development engineer believes that the lifetime in hours of a resistor follows an exponential distribution with a β parameter of 15,000. The probability that a resistor will last longer than 8,760 hours (1 year) is most nearly:

(A) 0.18
(B) 0.44
(C) 0.56
(D) 0.82

7. Consider a normal population with mean $\mu = 11$ from which the following random sample was obtained: $x_1 = 9$, $x_2 = 8.5$, $x_3 = 12$, $x_4 = 10$, and $x_5 = 10.5$. Based on this sample, the following statistics were calculated:

$$\bar{x} = 10,\ \sum_{i=1}^{5}(x_i - \bar{x})^2 = 7.5, \text{ and}$$

$$\sum_{i=1}^{5}(x_i - \mu)^2 = 12.5,$$

$$s^2 = \sum_{i=1}^{5}(x_i - \bar{x})^2/4 = 1.875$$

Assume that the variance σ^2 is unknown. It is desired to test the hypothesis that $\sigma^2 = 2.5$. The most precise statistic to test the hypothesis and its value calculated on the basis of the sample is:

(A) $z = \dfrac{\bar{x} - \mu}{\sigma} = -0.632$

(B) $t = \dfrac{\bar{x} - \mu}{s/\sqrt{n}} = -1.633$ with 4 degrees of freedom

(C) $\chi^2 = \dfrac{\sum_{i=1}^{5}(x_i - \mu)^2}{\sigma^2} = 5$ with 5 degrees of freedom

(D) $\chi^2 = \dfrac{\sum_{i=1}^{5}(x_i - \bar{x})^2}{\sigma^2} = 3$ with 4 degrees of freedom

GO ON TO THE NEXT PAGE

8. A 2^3 factorial experiment is run using the following levels. For X_1: 10 and 20; for X_2: 5 and 10; and for X_3: 20 and 30. The low level for each factor is represented by 1 and the high level is represented by 2. The following table shows the results obtained at random for the eight experimental conditions of the design:

X_1	X_2	X_3	Response
1	1	1	20
2	1	1	11
1	2	1	12
2	2	1	22
1	1	2	10
2	1	2	9
1	2	2	21
2	2	2	10

Which of the following individual choices would be most effective in increasing the value of the response in the region of the experimental conditions given above?

(A) Increase X3

(B) Increase X2

(C) Decrease X1

(D) Decrease X2

9. You are reading and processing data from a file using a program segment described below. The numbers in this file are 5, 2, 8, 4, 6, 9, 3, 7, 1, 0.

```
Set COUNT to 1 and A to 50
While COUNT <= 5
    Set J equal to COUNT
    Read the Jth number from the file and set A1 equal to it
    Increment COUNT by 1
Print A1
```

The function of this program segment is to:

(A) calculate the minimum of the numbers in the file

(B) calculate the maximum of the numbers in the file

(C) calculate the mean of the numbers in the file

(D) do none of the above

10. One of the most important performance criteria in the design of an information system (IS) is:

(A) how much data can be stored in the IS

(B) whether or not to computerize the IS

(C) how the user will use the information from the IS

(D) how much it costs to add/delete information records to/from the IS

11. A firm operates two different production lines, P and Q. Each production line produces the same two products, A and B, with the daily outputs as follows:

	Line P	Line Q
Product A	300	100
Product B	100	100

The firm is operating at considerably less than full capacity but has orders for 2,400 units of A and 1,600 units of B to be produced during the coming month. The costs of production are \$600/day for Line P and \$400/day for Line Q. Let x be the number of days that Line P is run and y be the number of days that Line Q is run. The goal is to determine the number of days to run each production line during the month in order to minimize monthly production costs. Which of the following models should be used?

(A) Minimize $600x + 400y$
subject to $300x + 100y \geq 2{,}400$
$100x + 100y \geq 1{,}600$
$x,y \geq 0$

(B) Minimize $600x + 400y$
subject to $300x + 100y \geq 1{,}600$
$100x + 100y \geq 2{,}400$
$x,y \geq 0$

(C) Minimize $600x + 400y$
subject to $300x + 100y \leq 2{,}400$
$100x + 100y \leq 1{,}600$
$x,y \geq 0$

(D) Minimize $400x + 600y$
subject to $300x + 100y \geq 2{,}400$
$100x + 100y \geq 1{,}600$
$x,y \geq 0$

12. Consider the following linear programming model:

$$\begin{aligned} \text{maximize } Z = \; & 5X_1 + 3X_2 & \\ \text{subject to } \; & X_1 + X_2 \leq 4 & \text{(Resource 1)} \\ & 2X_1 + 5X_2 \leq 10 & \text{(Resource 2)} \\ & X_1, X_2 \geq 0 & \end{aligned}$$

The increase in Resource 1 that will make its constraint redundant is most nearly:

(A) 0
(B) 1
(C) 2
(D) 6

13. For several years the manufacturing department of a company had operated with a high level of productivity, enthusiasm, and morale. People seemed to work fast and with a minimum of lost time. Productivity had been very good.

Recently, however, problems seemed to be developing that were causing errors that required work to be discarded or redone. The president felt that this was probably due to a breakdown in coordination or perhaps some personal conflicts. The president had some analyses made on the situation and reached several conclusions.

First, the corporate level managers were not communicating their goals and policies clearly to the operating level managers. In turn, these managers felt the top management was planning and communicating in a vacuum. They were making plans without relevant and realistic information from the operating levels. Lower-level managers in manufacturing and marketing frequently found themselves working at cross purposes and with little coordination from above. Constant bickering was evident between line and staff groups.

Second, key people at the operating levels frequently failed to "get the word" at all, or they got it indirectly through informal channels, perhaps too late.

Third, there were many cases of other functional departments at the operating level misunderstanding one another. At one time, the personnel department had performed a study for the sales department. However, when the sales department reviewed the study, it found that it was not really what it wanted and promptly filed the study away. Then one of the manufacturing sections felt that it needed some computer service, but what it got really did not do the job. There seemed to be a lack of common level of communication. The industrial engineers wanted an ergonomics evaluation for the production floor but were not able to obtain funding, even though several recent accidents causing back injuries had occurred.

Which type of managerial tactic should the president initiate to attempt to solve the problems?

(A) Participative management

(B) MBO

(C) Re-engineering

(D) TQM

GO ON TO THE NEXT PAGE

14. Consider the following precedence relationships for an activity network consisting of five activities a, b, c, d, and e:

1. a immediately precedes b
2. b immediately precedes c and d
3. c and d immediately precede e

The activity-on-arc CPM diagram (a dashed line represents a dummy activity with duration equal to zero) that best satisfies the precedence relationships is:

(A) 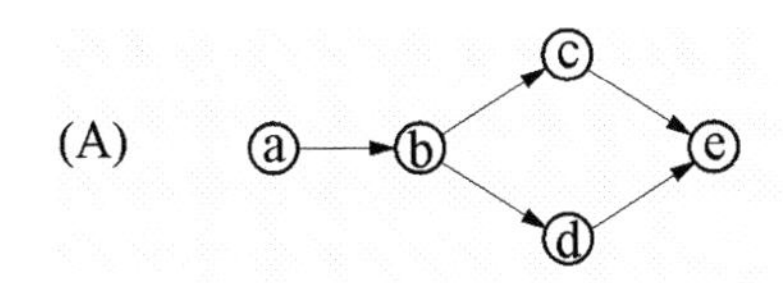

(B) 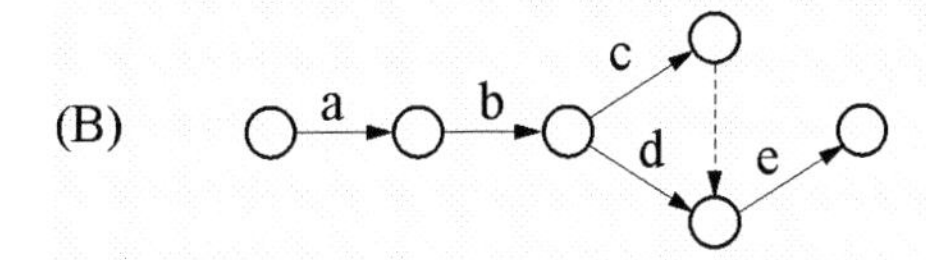

(C) 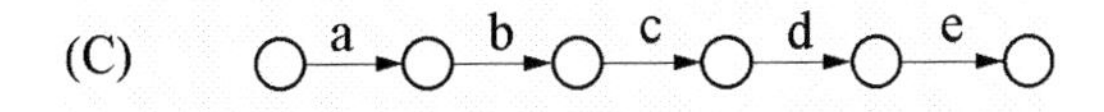

(D) 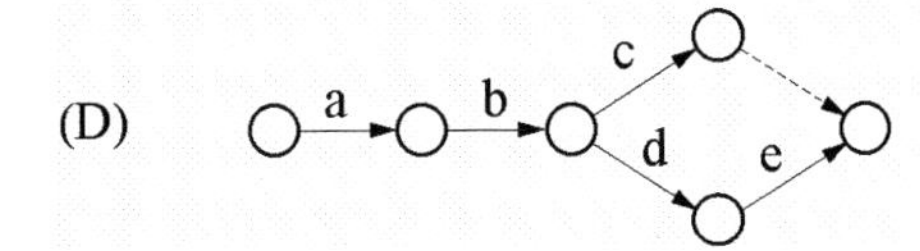

15. Which of the following best explains the partial CPM/PERT network shown below, where ϕ represents a dummy activity?

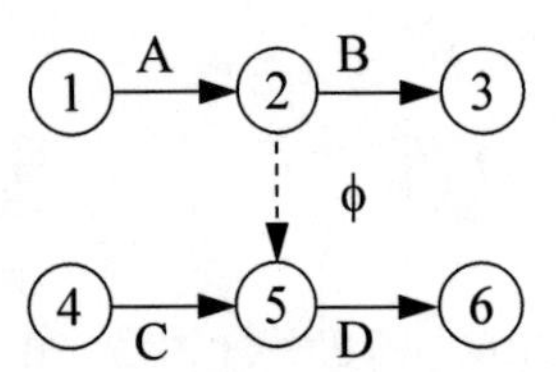

(A) Activity D cannot begin until both A and C are completed, but B can start after only A is completed.

(B) Activity B cannot begin until A is completed, and D cannot begin until C is completed.

(C) Activities B and D cannot begin until both A and C are completed.

(D) Activity B cannot begin until A, C, and D are completed.

16. Consider a flexible manufacturing system used to fabricate semiconductor wafers. Referring to the figure below, it consists of a load/unload chamber, an orient chamber (A), three identical process chambers (B, C, D), a cool-down chamber (E), and a robot that can move a single wafer at a time. Each of the Chambers A-E can process a single wafer at a time, and the load/unload station can hold up to 100 wafers. A batch of wafers is loaded into the load/unload station, and wafers are routed individually to Chamber A, then to **one** of the three chambers (B, C, or D) then to Chamber E, and back to the load/unload chamber. The figure shows deterministic process times. Assume the robot move times are negligible compared to process times. If more than one of Chambers B-D are available, the order of preference for moving a wafer from A is B, then C, then D.

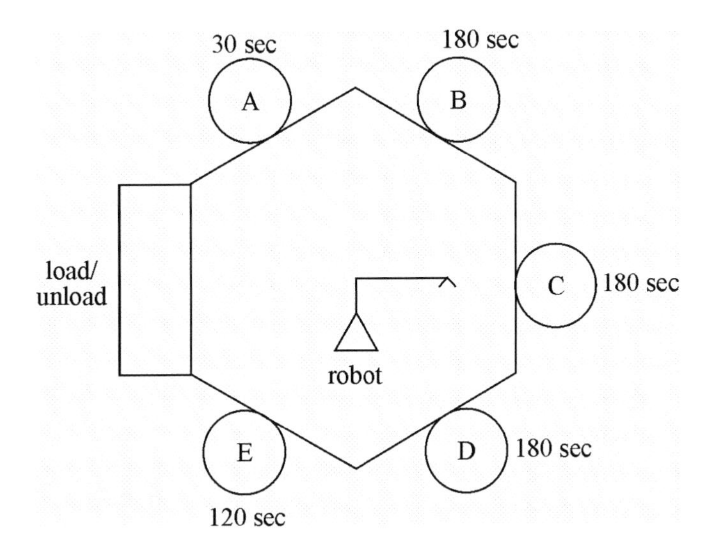

The bottleneck operation in this system is:

(A) Chamber A

(B) Chamber B

(C) Chamber E

(D) the robot

17. A production order calls for 2,000 pieces per week. There are enough machines of this type to meet the weekly requirement. Each machine used has a defective percent equal to 10%. Each machine used is available 80% of the time for processing this item. The total number of pieces, including scrap, that must be processed every week is most nearly:

(A) 2,000

(B) 2,200

(C) 2,223

(D) 2,500

18. In a flow shop, there are four parts that have to be processed on two machines, A and B, in that order. The following table shows the processing time of each part on each machine.

Part	Processing Time (min)	
	Machine A	Machine B
1	17	7
2	6	9
3	5	7
4	8	10

What is the shortest processing time (min) for these parts?

(A) 38

(B) 43

(C) 46

(D) 69

19. A manufacturer has five orders to fill, as shown below. Each of the five orders will be produced sequentially through three processes. The number of machines and monthly capacity of each machine are shown below.

Order	Quantity of Product
1	600
2	300
3	190
4	200
5	600

Process	Number of Machines	Monthly Capacity of Each Machine (units)
X	12	100
Y	8	250
Z	28	80

Assume that machines in each process are identical and that they do not fail. How many orders can be filled in one month's worth of production?

(A) 5
(B) 4
(C) 3
(D) 2

20. The main difference between a process or job-shop layout and a product layout is that in a product layout there are:

(A) very few products
(B) high volume product outputs
(C) high downtime costs if machines fail
(D) all of the above

GO ON TO THE NEXT PAGE

21. Consider the grid shown below.

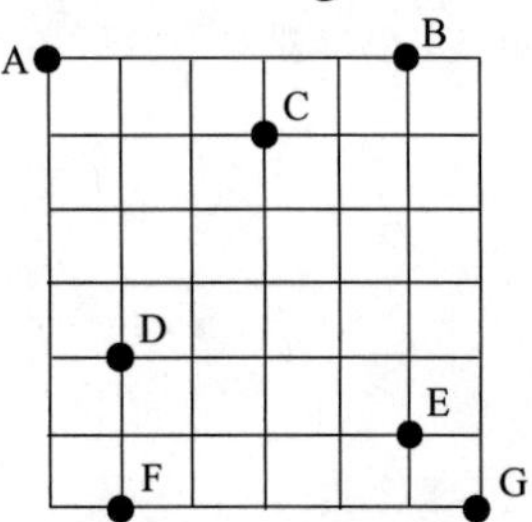

Using the rectilinear distance metric, which two points are equidistant from A?

(A) B and D

(B) B and E

(C) F and G

(D) No points are equidistant from A

22. A person-on-board order picker needs to pick the following items in order: A, B, C, D. The order picker consists of a crane that travels in an aisle. It travels using a rectilinear distance metric in the horizontal and vertical distances, and its speed is a constant one bay length per second. It starts at the home position and returns there only after picking all items. An actual pick operation consists of an insert (2 sec) and retrieval (2 sec). The items are located in the rack structure shown below.

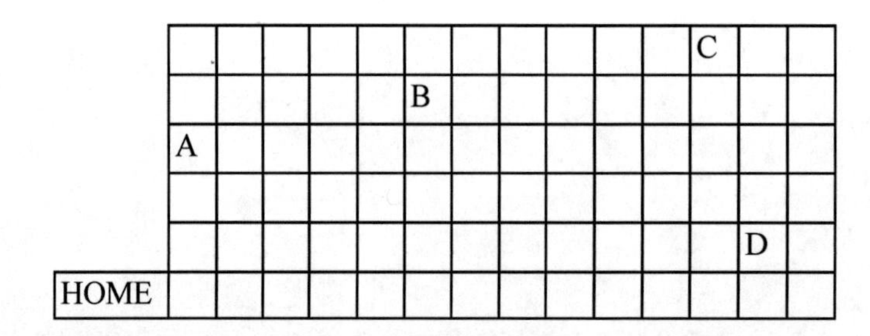

The total cycle time (sec) needed to pick all the items is most nearly:

(A) 36

(B) 44

(C) 47

(D) 52

23. A manufacturer of retail goods has one factory and three warehouses. The factory supplies all three warehouses, and the warehouses supply two retail stores as shown in the figure below:

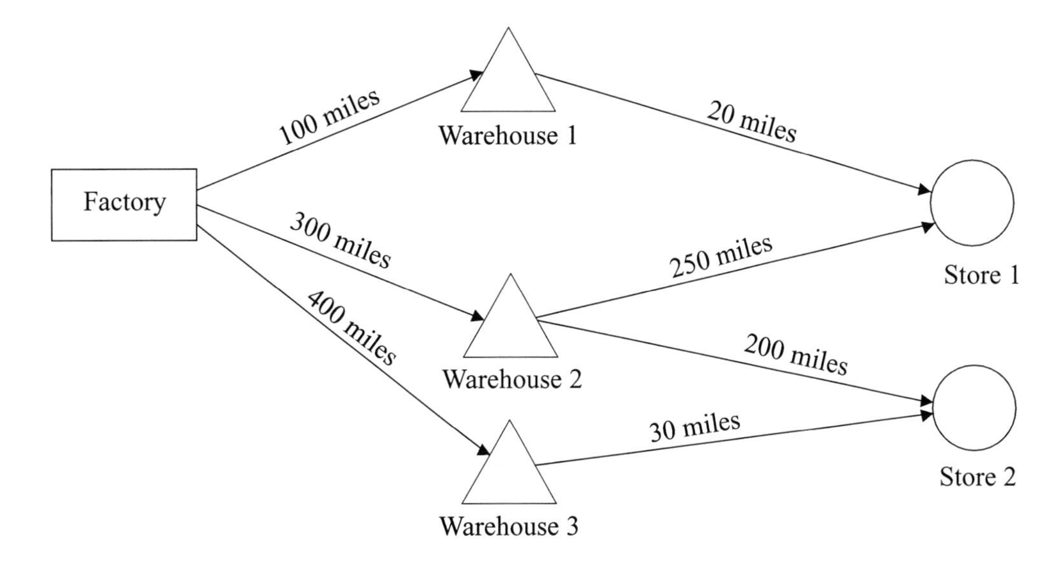

The manufacturer transports goods from the factory to the warehouses and then from the warehouses to the stores according to the following schedule and with the following costs:

From	To	Round Trips/Month	Cost/Mile
Factory	Warehouse 1	10	$1.00
Factory	Warehouse 2	4	1.00
Factory	Warehouse 3	8	1.00
Warehouse 1	Store 1	20	0.50
Warehouse 2	Store 1	10	0.50
Warehouse 2	Store 2	10	0.50
Warehouse 3	Store 2	20	0.50

The total monthly transportation cost is most nearly:

(A) $16,300

(B) $10,800

(C) $8,150

(D) $5,400

24. A time study of an assembly operation yielded the following results for a random sample consisting of ten observations:

Observation	Time (min)
1	8.25
2	8.20
3	8.28
4	8.29
5	8.40
6	8.35
7	8.20
8	8.16
9	8.20
10	8.40

The performance rating was established to be 110%. Using an allowance of 20% of the **job time**, the standard time (min) is most nearly:

(A) 8.3

(B) 9.0

(C) 10.9

(D) 11.4

25. You are predicting the time needed to perform a complicated procedure involving a machine. Each time an operation is performed, a person must prepare a part on the machine while the machine is not operating. Then the machine performs a complicated assembly and machining task. In the human part of this task, an 80% learning curve has been observed, with the time to do the 10th procedure observed to be 90 minutes and the machine operation taking an additional 240 min. The time (min) for these operations on the 30th unit is most nearly:

(A) 232

(B) 259

(C) 303

(D) 330

26. An operator is responsible for monitoring a number of critical sub-systems in a nuclear power plant. There are four computer screens to provide decision support for this responsibility. If a fault occurs in sub-system A, it could cause a serious safety problem in the plant. The operator must be notified immediately so as to render a report judgment in resolving the fault. Of the following, the best method used by the decision support system to notify the operator of such a fault is:

(A) an audible alarm

(B) a text warning message on a computer screen

(C) a graphical representation of an alarm condition on a computer screen

(D) an automatic system shutdown to terminate the fault

27. You are measuring the maximum capacity of a corridor for an emergency egress safety plan. An average human standing still in a crowd of people occupies approximately 0.20 m^2. The capacity (human occupants) of a corridor 100 m long × 3 m wide is most nearly:

(A) 39

(B) 60

(C) 300

(D) 1,500

GO ON TO THE NEXT PAGE

28. Total quality management (TQM) is now transforming American industry. Dr. W. Edwards Deming is called the father of this movement. His 14 points apply anywhere, to small organizations as well as to large ones, to the service industry as well as to manufacturing.

Which of the following is **NOT** one of the 14 points of TQM?

(A) Eliminate management by objective

(B) Create constancy of purpose toward improvement

(C) Eliminate work standards on the factory floor

(D) Depend on inspection to achieve quality

29. When Pareto analysis is used on a class of maintenance problems, the result is:

(A) the probability distribution of downtime for the average problem

(B) the mean time to repair the most frequent problem

(C) the variance of repair times

(D) a classification of problems within this class

30. You used the following data to develop $\bar{X}$ and R charts from samples with five observations each:

$\bar{X}$	R
15.7	3.6
18.3	4.7
17.6	4.2
16.2	3.8
17.7	4.6
17.5	3.9

Later, you took a sample of five observations and had the following values: 14.3, 16.7, 21.0, 21.2, 23.6. Based on all of these data, which of the following statements is most accurate?

(A) Both the mean and the range are in control.

(B) The mean is in control, but the range is not.

(C) The range is in control, but the mean is not.

(D) Neither the mean nor the range is in control.

IF YOU FINISH BEFORE TIME IS CALLED, YOU MAY WISH TO CHECK YOUR WORK ON THIS TEST

INDUSTRIAL
AFTERNOON SOLUTIONS

ANSWERS TO THE INDUSTRIAL AFTERNOON SAMPLE QUESTIONS

Detailed solutions to each question begin on the next page.

QUESTION	ANSWER
1	B
2	D
3	A
4	C
5	D
6	C
7	C
8	B
9	D
10	C
11	A
12	B
13	A
14	B
15	A
16	C
17	C
18	B
19	C
20	D
21	A
22	D
23	A
24	C
25	C
26	A
27	D
28	D
29	D
30	B

INDUSTRIAL AFTERNOON SOLUTIONS

1. Investment = ($25,000)(1.00) = –$25,000

Salvage Value (P/F value @ 10%) = ($5,000)(0.6209) = $3,104.50

Total Cost Savings = 0.25 × 45,000 – 0.40 × 2,500 = 10,250

P/A @ 10% = (10,250)(3.7908) = 38,855.70

Σ = –25,000 + 3,104,50 + 38,855.70 = $16,960.20

THE CORRECT ANSWER IS: (B)

2. $$\text{Labor cost/truck} = \frac{9 \text{ min}}{60 \text{ min/hr}} \times 10.00/\text{hr} = \$1.50$$

Materials cost/truck = $1.50

Net profit per truck = $12.00

$$\text{Breakeven} = \frac{(\$1{,}200 + \$300 + \$1{,}500)}{\$12.00/\text{truck}} = 250 \text{ trucks}$$

THE CORRECT ANSWER IS: (D)

3. C_{pr} = non-productive cost + machining cost + tool change cost + cost of tools

THE CORRECT ANSWER IS: (A)

4. Capitalized cost = first cost + annual cost of recurring expenses/interest rate

= 800,000 + (75,000 + 300,000 (A/F, 0.08, 5)/0.08
= 800,000 + (75,000 + 300,000 × 0.1705)/0.08
= 800,000 + 126,150/0.08
= 2,376,875

THE CORRECT ANSWER IS: (C)

5. It is assumed that the dimensions have a normal distribution.

Then, the answer is $P(Z > ((6.80 - 6.55)/0.15)) + P(Z < ((6.45 - 6.55)/0.15)$

$= P(Z > 1.67) + P(Z < -0.67)$

$= (0.0548 - 0.007) + (0.2743 - 0.7 \times 0.0323)$

$= 0.047 + 0.252$

$= 0.299$

THE CORRECT ANSWER IS: (D)

6. $P(\text{Life} > 1 \text{ yr}) = 1 - P(\text{Life} \leq 1 \text{ yr}) = 1 - F_T(8{,}760) = e^{-8{,}760/15{,}000} = 0.5577$

THE CORRECT ANSWER IS: (C)

7. $\chi^2 = \dfrac{\sum_{i=1}^{n}(x_i - \mu)^2}{\sigma^2}$ follows a chi-square distribution with n degrees of freedom

$\chi^2 = \dfrac{12.5}{2.5} = 5.0$ Here $n = 5$ degrees of freedom.

THE CORRECT ANSWER IS: (C)

8. First find the effects of the three factors then interpret these effects as follows:

(a) a positive effect means that the response is increased if the magnitude of the factor is increased.

(b) a negative effect means that the response is increased if the magnitude of the factor is decreased

$E_1 = \frac{1}{4}(-20+11-12+22-10+9-21+10) = -2.75$

$E_2 = \frac{1}{4}(-20-11+12+22-10-9+21+10) = 3.75$

$E_3 = \frac{1}{4}(-20-11-12-22+10+9+21+10) = -3.75$

THE CORRECT ANSWER IS: (B)

9. The program segment prints the first five numbers in the file.

THE CORRECT ANSWER IS (D)

10. Choice (C) is the most important criterion in the design of any information, because if one does not consider how the user will use the information, then the purpose for collecting the information in the first place is unwarranted.

THE CORRECT ANSWER IS: (C)

11. The monthly production cost would be $600x + 400y$. For these production times the firm would produce $300x + 100y$ units of A and $100x + 100y$ units of B. However, because of the order requirements for the month, we must have $300x + 100y \geq 2{,}400$ and $100x + 100y \geq 1{,}600$.

THE CORRECT ANSWER IS: (A)

12. When the line for Resource 1 passes through the point (5,0), the constraint becomes redundant. Thus, the increase is $5 - 4 = 1$

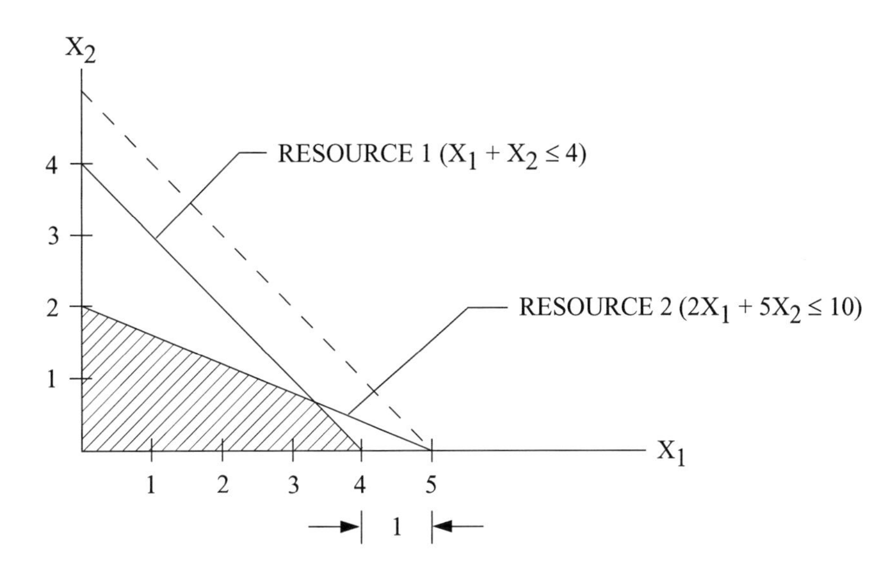

THE CORRECT ANSWER IS: (B)

13. Allowing employees at all levels to become involved in decision making can help the situation at this company.

THE CORRECT ANSWER IS: (A)

14. A dummy arc is used to avoid two arcs joining the same two nodes to represent the third precedence relationship. In this case, either of the activities a or b can be followed by a dummy arc before joining the first node of activity e.

THE CORRECT ANSWER IS: (B)

15. The dummy activity linking events 2 and 5 indicates that D cannot begin until A is completed. Also, the figure shows that D cannot begin until C is completed, and B does not depend on C.

THE CORRECT ANSWER IS: (A)

16. Chamber E is the bottleneck operation. This system can be thought of as a serial system with three process steps. Process step 1 (orient) has a production rate of 1 wafer/30 sec. Process step 2 (B–D) has a production rate of 1 wafer/60 sec since there are three processors. Process step 3 (E) has a production rate of 1 wafer/120 sec, and is therefore the bottleneck.

THE CORRECT ANSWER IS: (C)

17. $\frac{2,000}{0.90} = 2,222.22$

THE CORRECT ANSWER IS: (C)

18. Sort the jobs by the processing times.

Part	Lathe	Drilling
3	5	7
2	6	9
1	17	7
4	8	10

Schedule the shortest processing time by the following rules. If the shortest processing time appears on the first machine, schedule that job to be processed first on the list of the first machine (A). If it appears on the second machine, schedule that job to be the last one on the second machine (B). Delete that job, and continue until all the jobs are scheduled.

The sequence is:

Part	Machine A	Machine B
3	5	7
2	6	9
4	8	10
1	17	7

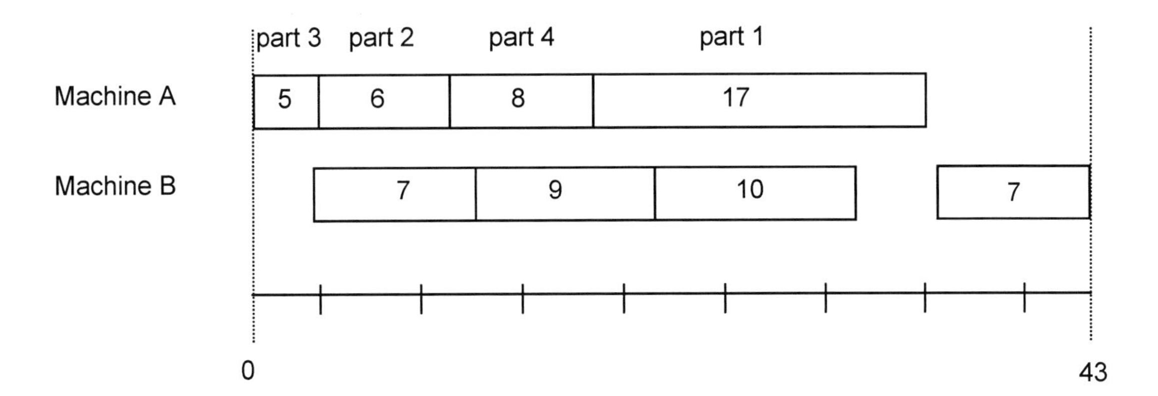

Thus, the processing time is 43 minutes.

THE CORRECT ANSWER IS: (B)

19. The capacity of the system is the minimum of the capacities of the process.

= min[(12 machines)(100 units/machine), (8)(250), (28)(80)]

= min(1,200, 2,000, 2,240) = 1,200

Orders must be processed according to the order in the table. Therefore, only the first <u>three</u> orders can be filled in this month's production.

Cumulative of first three orders = 1,090
Cumulative of first four orders = 1,290 (infeasible)

THE CORRECT ANSWER IS: (C)

20. The factors mentioned in (A) (B) and (C) are all characteristics of a product layout and no single one of them can be excluded.

THE CORRECT ANSWER IS: (D)

21. Points B and D are both five units distant from A under the rectilinear metric. The distances of the remaining points are as follows:

C	4
E	10
F	7
G	12

THE CORRECT ANSWER IS: (A)

22. Cycle time = Travel (HOME, A) + Pick + Travel (A, B) + Pick + Travel (B, C) + Pick + Travel (C, D) + Pick + Travel (D, HOME)

Cycle time = 4 + (2 + 2) + 6 + (2 + 2) + 7 (2 + 2) + 5 (2 + 2) + 14 = 52

THE CORRECT ANSWER IS: (D)

23. Cost is calculated below:

Factory to WH 1	=	(2)(100 mi/trip)[10 trips/mo($1/mi)]	=	$2,000
WH 2	=	(2)(300)(4)($1)	=	2,400
WH 3	=	(2)(400)(8)($1)	=	6,400
WH1 to S1	=	(2)(20)(20)($0.50)	=	400
WH2 to S1	=	(2)(250)(10)($0.50)	=	2,500
WH2 to S2	=	(2)(200)(10)($0.50)	=	2,000
WH3 to S3	=	(2)(30)(20)($0.50)	=	600
		Total Cost	=	$16,300

THE CORRECT ANSWER IS: (A)

24. The average of the ten observations is 8.273 min. This observed time must be multiplied by the performance factor (1.10) to obtain the normal time (NT). That is, NT = 8.273 × 1.10 = 9.1003. Now, this normal time must be multiplied by the allowance factor (AF) to get the standard time. ST = NT × AF. The allowance factor in this problem is computed as $AF = AF_{job} = 1 + A$, where A = allowance fraction based on job time, that is A = 0.20. Therefore, $AF_{day} = 1 + 0.20 = 1.2$. Then, ST = 1.20 × 9.1003 = 10.92 min.

THE CORRECT ANSWER IS: (C)

25. Total time = operator time + machine time
To calculate operator time, $t(10) = 90 = K \times 10^{(\ln 0.80/\ln 2)}$

$$= K \times 10^{-0.3219} = K \times 0.4765$$

$$\text{So } K = 188.86$$

Operator time = $188.86 \times 30^{-0.3219} = 63.2$ min
Machine time doesn't change, so total time = 240 + 63.2 = 303.2 min

THE CORRECT ANSWER IS: (C)

26. Choice (A) is correct since it provides the most immediate notification, and the operator can start working to resolve the fault. Choices (B) and (C) are not appropriate since the operator may not be looking at the right computer screen at the time of a fault. Choice (D) is not appropriate since the operator is tasked with resolving the fault.

THE CORRECT ANSWER IS: (A)

27. Capacity = Area of corridor/Area per occupant
= (100 × 3)/0.2
= 1,500 occupants

THE CORRECT ANSWER IS: (D)

28. Option (D) is the correct answer since it is NOT one of Dr. Deming's 14 points for TQM. It is exactly opposite to his point number 3.

THE CORRECT ANSWER IS: (D)

29. Pareto analysis is used to determine which problems are most significant using some criterion, usually frequency of occurrence.

THE CORRECT ANSWER IS: (D)

30. From the data used to develop the correct charts, $\bar{\bar{x}} = 17.17$ and $\bar{R} = 4.13$. The control limits for n = 5 are:

$UCL_{\bar{x}} = 17.17 + 0.577 \times 4.13 = 19.55$

$LCL_{\bar{x}} = 17.17 - 0.577 \times 4.13 = 14.79$

$LCL_R = 0$

$UCL_R = 2.114 \times 4.13 = 8.73$

For the given observations, $\bar{X} = 19.36$, $R = 9.3$

THE CORRECT ANSWER IS: (B)

FE Study Material Available for Purchase

FE Supplied-Reference Handbook

Sample Questions and Solutions (in printed and a multidiscipline, CD-ROM format)
are available for the following modules:
Chemical
Civil
Electrical
Environmental
General
Industrial
Mechanical

For more information about these and other Council publications and services,
visit us at www.ncees.org or contact our
Customer Service Department at (800) 250-3196.